Shafarevich Maps and
Automorphic Forms

M. B. PORTER LECTURES

RICE UNIVERSITY, DEPARTMENT OF MATHEMATICS

SALOMON BOCHNER, FOUNDING EDITOR

Recurrence in Ergodic Theory and Combinatorial Number Theory,
by H. Furstenberg (1981)

The Geometry and Dynamics of Magnetic Monopoles,
by M. Atiyah and N. Hitchin (1988)

Topics in Noncommutative Geometry,
by Y. I. Manin (1991)

Shafarevich Maps and Automorphic Forms,
by J. Kollár (1995)

Shafarevich Maps and Automorphic Forms

JÁNOS KOLLÁR

PRINCETON UNIVERSITY PRESS
PRINCETON, NEW JERSEY

Published by Princeton University Press, 41 William Street,
Princeton, New Jersey 08540
In the United Kingdom Princeton University Press, Chichester,
West Sussex

Library of Congress Cataloging-in-Publication Data

Kollar, Janos
Shafarevich maps and automorphic forms / Janos Kollar
p cm
Includes bibliographical references and index
ISBN 0-691-04381-7
1 Automorphic forms 2 Shafarevich maps 3 Complex manifolds
I Title
QA331 K728 1995
515' 9--dc20 94-46270

This book has been composed in Times Roman

Princeton University Press books are printed on acid-free paper and meet the guidelines for permanence and durability of the Committee on Production Guidelines for Book Longevity of the Council on Library Resources

Printed in the United States of America

10 9 8 7 6 5 4 3 2 1

CONTENTS

Preface vii
Acknowledgments ix
Introduction 3

PART I. SHAFAREVICH MAPS

CHAPTER 1. Lefschetz-Type Theorems for π_1 19
CHAPTER 2. Families of Algebraic Cycles 27
CHAPTER 3. Shafarevich Maps and Variants 36
CHAPTER 4. The Fundamental Group and the Classification of Algebraic Varieties 49

PART II. AUTOMORPHIC FORMS: CLASSICAL THEORY

CHAPTER 5. The Method of Poincaré 59
CHAPTER 6. The Method of Atiyah 71
CHAPTER 7. Surjectivity of the Poincaré Map 81
CHAPTER 8. Ball Quotients 92

PART III. VANISHING THEOREMS

CHAPTER 9. The Kodaira Vanishing Theorem 105
CHAPTER 10. Generalizations of the Kodaira Vanishing Theorem 115
CHAPTER 11. Vanishing of L^2-Cohomologies 127
CHAPTER 12. Rational Singularities and Hodge Theory 133

PART IV. AUTOMORPHIC FORMS REVISITED

CHAPTER 13. The Method of Gromov 141
CHAPTER 14. Nonvanishing Theorems 151
CHAPTER 15. Plurigenera in Etale Covers 161
CHAPTER 16. Existence of Automorphic Forms 167

PART V. OTHER APPLICATIONS AND SPECULATIONS

CHAPTER 17. Applications to Abelian Varieties 175
CHAPTER 18. Open Problems and Further Remarks 183

References 191
Index 201

PREFACE

The theory of automorphic forms goes back to L. Euler, but the main development of its function-theoretic aspects started with H. Poincaré. The aim of the theory is to study the function theory and geometry of complex manifolds whose universal covering space is well understood, for instance $\mathbb{C}^n$ or the unit ball in $\mathbb{C}^n$.

Around 1970, I. R. Shafarevich conjectured that the universal cover of a smooth projective algebraic variety is holomorphically convex. This question is still completely open. The Shafarevich conjecture suggests that the most general version of the theory of automorphic forms is the study of projective varieties whose universal covering space is Stein. The function theory of Stein spaces is quite well understood; therefore, one hopes that there are many interesting connections between the meromorphic function theory of a variety and the holomorphic function theory of its universal cover.

The aim of these notes is to connect a weakened version of the Shafarevich conjecture with the theory of automorphic forms, thereby providing general methods to study varieties whose fundamental group is large. Instead of attempting to prove the Shafarevich conjecture, I try to concentrate on some ideas that connect the fundamental group directly to algebro-geometric properties of a variety. Some of the results are rather promising but much more remains to be done.

These notes are a considerably expanded version of the Milton Brockett Porter lectures I delivered at Rice University in March 1993. I would like to thank the Mathematics Department of Rice University for its warm hospitality and for giving me the first occasion to test my theories on a large audience.

I had the opportunity to give further series of lectures about these topics at the University of Bayreuth, the University of Utah, and at the Regional Geometry Institute at Park City, Utah. I am grateful to my audiences for pointing out several mistakes and suggesting improvements in the presentation.

ACKNOWLEDGMENTS

I discussed various parts of these notes with J. Carlson, S. Gersten, J. Johnson, L. Katzarkov, R. Lazarsfeld, L. Lempert, C. McMullen, S. Mori, A. Moriwaki, R. Narasimhan, M. Nori, H. Rossi, J. Taylor, D. Toledo. They helped me considerably with various questions.

The comments of S. Kovács, R. Mayer, E. Szabó, and J. Winkelman were especially crucial.

Partial financial support was provided by the National Science Foundation under grant numbers DMS-8707320 and DMS-9102866, and by the Monbusho International Science Program, number 04044081.

These notes were typeset in $\mathcal{A}\mathcal{M}\mathcal{S}$-TEX, the TEXmacro system of the American Mathematical Society.

Shafarevich Maps and Automorphic Forms

INTRODUCTION

The aim of these notes is to study algebraic varieties whose fundamental group is nontrivial.

Fundamental groups of algebraic varieties have been studied quite extensively. Most of the research has centered on trying to determine which groups occur as fundamental groups of algebraic varieties. This problem is very interesting, and there have been numerous important results. [Arapura94] is a recent overview of this direction.

My main interest is to see how the presence of a "large" fundamental group influences other algebro-geometric properties of a variety. The main examples I have in mind are the Kodaira dimension and the pluricanonical maps.

So far the methods of these two approaches have very little in common, and the results also seem rather independent. I hope that in the future the two directions will converge, giving a much better understanding of both questions.

One of the motivations for my approach comes from homotopy theory. The following summary concentrates on those ideas that are important for the algebro-geometric version.

The Topological Theory

In homotopy theory the customary method for analyzing the effect of the fundamental group on a topological space has two steps.

0.0.1 Identify those topological spaces where $\pi_1(X)$ determines everything. Let G be a discrete group. Up to homotopy, there is a unique CW-complex $K(G,1)$ such that $\pi_1(K(G,1)) = G$ and $\pi_i(K(G,1)) = 0$ for $i \geq 2$. The space $K(G,1)$ is called the *Eilenberg-MacLane space* of G.

This definition is rather difficult to reformulate for algebraic varieties for lack of a good algebraic version of the higher homotopy groups. Fortunately, there is an equivalent characterization of Eilenberg-MacLane spaces that involves only the fundamental group:

THEOREM. *Let X be a CW-complex. The following are equivalent:*

(0.0.1.1) $\pi_i(X) = 0$ for every $i > 0$.

(0.0.1.2) If Y is any CW-complex and $f : Y \to X$ is continuous, then f is null homotopic iff $f_* : \pi_1(Y) \to \pi_1(X)$ is the trivial homomorphism.

The second variant can easily be transformed into a meaningful definition in algebraic geometry (0.3.1).

0.0.2 Classifying maps. Let X be any CW-complex. Then there is a map $f : X \to K(\pi_1(X), 1)$ such that $f_* : \pi_1(X) \to \pi_1(K(\pi_1(X), 1)) = \pi_1(X)$ is the identity. The map f is unique up to homotopy. Furthermore, in the homotopy category X can be viewed as a fiber bundle over $K(\pi_1(X), 1)$ with simply connected fiber.

In order to point out the principal directions of the algebro-geometric theory, I first outline the main steps of the Abelian version, namely those that describe the influence of the first homology group on the structure of a variety.

The Abelian Theory

Let X be a smooth projective variety over $\mathbb{C}$. What can one say about X if we know something about $H_1(X, \mathbb{Z})$? The traditional approach can be divided into several steps.

0.1.1 Identify those varieties where $H_1(X, \mathbb{Z})$ determines everything. Let $L \subset \mathbb{C}^n$ be a discrete Abelian subgroup of rank $2n$. $\mathbb{C}^n/L$ is a compact complex manifold called a *complex torus*. Under suitable conditions (see, e.g., [Mumford68]) the quotient is an algebraic variety, called an *Abelian variety*.

$\mathbb{C}^n/L$ is uniquely determined up to isomorphism by $H_1(\mathbb{C}^n/L, \mathbb{Z})$ together with its Hodge structure.

0.1.2 Albanese morphism. If X is a compact Kähler manifold (or a smooth projective variety), then there is a unique complex torus $\mathrm{Alb}(X) = \mathbb{C}^n/L$ and a morphism

$$\mathrm{alb}_X : X \to \mathrm{Alb}(X),$$

which induces an isomorphism $H_1(X, \mathbb{Z})/(\text{torsion}) \cong H_1(\mathrm{Alb}(X), \mathbb{Z})$. $\mathrm{Alb}(X)$ is called the *Albanese variety* of X, and alb_X the *Albanese morphism*.

The precise definition of alb_X is not very important right now; the following is meant only as a quick reminder. Fix a basis $\omega_1, \ldots, \omega_k \in H^0(X, \Omega_X^1)$ and a point $x_0 \in X$. Define

$$\mathrm{alb}_X : X \to \mathrm{Alb}(X) \quad \text{by} \quad \mathrm{alb}_X(x) = \left(\int_{x_0}^{x} \omega_1, \ldots, \int_{x_0}^{x} \omega_k\right).$$

Holomorphic global 1-forms are closed; therefore, the integral depends only on the homology class of the path from x_0 to x. Thus the natural target of alb_X is

$$\mathrm{Alb}(X) := H^0(X, \Omega^1_X)^*/H_1(X, \mathbb{Z}).$$

0.1.3 The corresponding fiber space. The Albanese morphism can be factored as

$$\mathrm{alb}_X : X \xrightarrow{a_X} A(X) \xrightarrow{a'_X} \mathrm{Alb}(X),$$

where a_X has connected fibers and $a'_X : A(X) \longrightarrow \mathrm{Alb}(X)$ has finite fibers. One sees that $a'_X = \mathrm{alb}_{A(X)}$. Thus the study of an arbitrary variety can be done in two steps:

(i) study varieties Y such that alb_Y has finite fibers;

(ii) study X as a fiberspace over $A(X)$, using induction on the dimension to understand the fibers of a_X.

Unfortunately $A(X)$ can have arbitrary normal singularities. Therefore it is frequently more convenient to resolve the singularities of $A(X)$. Correspondingly, in the above first step we need to study smooth varieties Y such that alb_Y has finite fibers over a dense open set. (Such a morphism is called *generically finite*.)

It is useful to note that a_X can be defined intrinsically in several ways. For our purposes the following version is the most useful:

Let $Z \subset X$ be an irreducible subvariety. Then

$$a_X(Z) = \text{point} \quad \Leftrightarrow \quad \mathrm{im}[H_1(Z, \mathbb{Z}) \to H_1(X, \mathbb{Z})] \quad \text{is finite.}$$

This in turn yields the following characterization:

$$a_X \text{ is finite} \quad \Leftrightarrow \quad \begin{array}{l} \mathrm{im}[H_1(Z, \mathbb{Z}) \to H_1(X, \mathbb{Z})] \text{ is infinite} \\ \forall\, Z \subset X,\ \dim Z > 0. \end{array}$$

0.1.4 Function theory of abelian varieties. Abelian varieties can be studied very thoroughly using *theta-functions* (see, e.g., [Mumford68; Siegel73]). Theta-functions are holomorphic functions on the universal covering of an Abelian variety that are "close" to being periodic mod L. There is a one-to-one correspondence between theta-functions on $\mathbb{C}^n$ and sections of line bundles on $\mathbb{C}^n/L$.

However, from the general point of view, theta-functions are too special. Elementary properties of theta-functions translate into very difficult questions in the general theory.

0.1.5 Varieties generically finite over an Abelian variety. If alb_Y is generically finite, then Y has several very special and pleasant properties. Some of these are summarized next. For the proofs and for further results, see (17.8–12).

THEOREM. *Let Y be a smooth projective variety and assume that* alb_Y *is generically finite. Then*
(0.1.5.1) $H^0(X, K_X) \geq 1$, *and*
(0.1.5.2) [Green-Lazarsfeld87] $\chi(X, K_X) \geq 0$.

0.1.6 Applications to pluricanonical maps. This direction has motivated a lot of research, starting with the results of [Ueno75]. One case where optimal results are known is the following:

THEOREM. [Kollár86a, 6.2, Kollár93b, 10.4] *Let X be a smooth projective threefold of general type such that* $\operatorname{rank} H^1(X, \mathbb{Z}) > 0$.
Then $H^0(X, K_X^2) \geq 1$.

It is a very interesting problem to develop an analogous theory based on the fundamental group instead of the first homology group. At the moment only partial results are known, in some cases even the right conjectures are unclear. In order to make the program more explicit, I start with the description of a very special case where most of the answers are known.

The Non-Abelian Theory: Ball Quotients

Fix a group $\Gamma < SU(1, n)$ which acts on the unit ball $B^n \subset \mathbb{C}^n$ without fixed points such that the quotient $\Gamma\backslash B^n$ is compact. Γ is called a *surface group* if $n = 1$ (i.e., Γ is the fundamental group of a Riemann surface of genus at least 2).

0.2.1 Identify those varieties where $\pi_1(X) = \Gamma$ determines everything. By [Yau77], if X is a smooth projective variety such that $\pi_1(X) \cong \Gamma$ and the universal cover of X is contractible, then X itself is a ball quotient. Thus again we find that the topology very much determines the algebraic structure.

0.2.2 Morphisms to ball quotients. More generally, we could ask the following. Let Y be a smooth projective variety. Given a quotient of the fundamental group $\sigma : \pi_1(Y) \twoheadrightarrow \Gamma$, under what conditions does there exist a morphism $f(\sigma) : Y \to \Gamma\backslash B^n$ which induces σ on π_1? (Even if Π is a free Abelian group, a homomorphism $\pi_1(Y) \to \Pi$ usually does not correspond to a morphism $Y \to$ (Abelian variety), unless Π is the maximal Abelian quotient. Similar maximality assumptions are always necessary.)

There are two cases to distinguish:

(0.2.2.1) Γ is a *surface group.*

As in the Abelian case assume that σ cannot be factored through a larger genus surface group. Then by [Siu87, 4.7] there is a morphism $f(\sigma) : X \to C$ onto a smooth curve such that

$$[f(\sigma)_* : \pi_1(X) \to \pi_1(C)] \cong [\sigma : \pi_1(X) \to \Gamma].$$

(0.2.2.2) Γ acts on B^n, $n \geq 2$. (Up to isomorphism there are two different actions of Γ; they are complex conjugates of each other. In what follows we always have to choose the suitable one.)

The situation in this case is even better. We need the maximality assumption that σ cannot be factored through a surface group. The results of [Siu80, Carlson-Toledo89] imply that there is a unique morphism $f(\sigma) : X \to \Gamma\backslash B^n$ such that

$$[f(\sigma)_* : \pi_1(X) \to \pi_1(\Gamma\backslash B^n)] \cong [\sigma : \pi_1(X) \to \Gamma].$$

0.2.3 The Corresponding fiber space. As in the Abelian case, the above morphism $f(\sigma) : X \to \Gamma\backslash B^n$ can be factored as

$$f(\sigma) : X \longrightarrow S(X) \xrightarrow{\sigma_X} \Gamma\backslash B^n,$$

where σ_X is finite. It is frequently more convenient to take a smooth model of $S(X)$ and study those varieties that admit a generically finite morphism to a ball quotient.

0.2.4 Automorphic forms. Γ-automorphic forms on B^n are holomorphic functions on B^n that transform in a "simple" way under the action of Γ. They can be identified with sections of some power of the canonical line bundle $K_{\Gamma\backslash B^n}$. Thus the theory of automorphic forms is the study of pluricanonical maps of ball quotients.

In the Abelian case it is easy to write down theta-functions explicitly. In the ball quotient case this is much harder to do. A general method is given by the theory of Poincaré series. It is, however, not a priori clear that the Poincaré series converge, or that they yield all automorphic forms.

These problems have been intensively studied, starting with the case where B is the unit disc in $\mathbb{C}$ [Poincaré1883]. Later these results were extended to the case of bounded symmetric domains in full generality, and to bounded nonsymmetric domains in $\mathbb{C}^n$ to a lesser extent [Siegel73]. Some of these results are discussed in chapters 5 to 8.

One of the basic results is formulated at the end of (0.2.5).

0.2.5 Varieties generically finite over a ball quotient. This property can be defined intrinsically as follows.

Definition-Proposition. Let X be a normal projective variety. The following two properties are equivalent:
(0.2.5.1) X admits a finite morphism $f : X \to \Gamma\backslash B^n$.
(0.2.5.2) There is a homomorphism $\pi_1(X) \to \Gamma$ such that

$$\operatorname{im}[\pi_1(Z) \to \pi_1(X) \to \Gamma] \quad \text{is infinite}$$
$$\text{for all } Z \subset X, \dim Z > 0.$$

As before, it is technically easier to study the following related property.

Definition-Proposition. Let X be a normal projective variety. The following two properties are equivalent:
(0.2.5.1.gen) X admits a generically finite morphism $f : X \to \Gamma\backslash B^n$.
(0.2.5.2.gen) There is a homomorphism $\pi_1(X) \to \Gamma$ such that

$$\operatorname{im}[\pi_1(Z) \to \pi_1(X) \to \Gamma] \quad \text{is infinite}$$
$$\text{for “most” } Z \subset X, \dim Z > 0.$$

Several of the results mentioned in (0.1.5) have their non-Abelian counterparts. These follow from the more general versions given in (6.5) and (5.22).

THEOREM. *Let X be a smooth projective variety that admits a generically finite morphism $f : X \to \Gamma\backslash B^n$. Then*
(0.2.5.3) $H^0(X, K_X^2) \geq 1$, and
(0.2.5.4) $H^0(X, K_X^m)$ defines a birational map for $m \gg 1$.

The Non-Abelian Theory: General Case

0.3.1 Identify those varieties where $\pi_1(X)$ determines everything. This seems very difficult, maybe even hopeless.

From the topological point of view a good question is to understand those algebraic varieties which are Eilenberg-MacLane spaces. This is the strongest variant of the problem.

Assume that Π occurs as the fundamental group of an algebraic variety Y. Does there exist a variety $U(\Pi)$ such that $\pi_1(U(\Pi)) \cong \Pi$ and the universal cover of $U(\Pi)$ is contractible? Can one describe all such varieties? Is there a morphism $Y \to U(\Pi)$ that induces an isomorphism on π_1?

These questions need to be refined. First of all, a finite group Π does not have any finite dimensional Eilenberg-MacLane space. Thus, the best we can hope for is a group Π' such that $X(\Pi')$ exists and Π' is "commensurable" with Π (e.g., in the sense that there is a homomorphism $\Pi \to \Pi'$ with finite kernel and cokernel).

Even with this weaker version there are some further problems. [Toledo90] observed that if Π is a noncocompact lattice in $SU(1, n)$, then Π has odd cohomological dimension (thus $U(\Pi)$ does not exist), but there are smooth projective varieties whose fundamental group is Π. In these cases there is a quasi-projective variety that plays the role of $U(\Pi)$.

The classifying space of Hodge structures suggests that in some cases $U(\Pi)$ may exist only as a complex manifold that does not admit any algebraic structure.

Unfortunately, I am unable to say anything about this very interesting direction.

Another approach is to try to find an algebro-geometric analog of Eilenberg-MacLane spaces. This is made possible by the second characterization (0.0.1.2). We can change this to an algebro-geometric definition by requiring the property only for morphisms of algebraic varieties. Since a nonconstant morphism between proper algebraic varieties is never null homotopic (not even after passing to finite covers), the following is the precise analog of (0.0.1.2):

Definition. Let X be a proper variety. We say that X has *large fundamental group* if

$$\operatorname{im}[\pi_1(Y) \to \pi_1(X)] \quad \text{is infinite}$$
$$\text{for every nonconstant morphism } Y \to X.$$

At least conjecturally, these spaces have a very nice characterization.

SHAFAREVICH CONJECTURE. [Shafarevich72, IX.4.3] *Let X be a normal and proper variety. The following are equivalent:*
(0.3.1.1) X has large fundamental group.
(0.3.1.2) The universal cover of X is Stein.

(0.3.1.2) $\Rightarrow$ (0.3.1.1) is elementary but the converse seems very difficult.

0.3.2 Morphisms to $U(\Pi)$. More generally, we could ask the following: Given a quotient of the fundamental group $\sigma : \pi_1(Y) \twoheadrightarrow \Pi$, under what conditions does there exist a morphism $f(\sigma) : Y \to U(\Pi)$ that induces σ on π_1?

For many of the cases when $U(\Pi)$ is known to exist, the answer to this more general question is also known.

Examples. (0.3.2.1) Π is Abelian. This corresponds to the Albanese morphism.

(0.3.2.2) Π is a *cocompact lattice* in $SU(1, n)$ acting freely on the complex n-ball B^n. This was discussed in (0.2.2).

(0.3.2.3) Abelian by finite groups.

Π is called *Abelian by finite* if it has a finite index Abelian subgroup. Assume for simplicity that $\pi_1(X)$ is Abelian by finite. Let $H \triangleleft \pi_1(X)$ be an Abelian normal subgroup and set $G = \pi_1(X)/H$. Let $X' \to X$ be the corresponding Galois cover with Galois group G. $\pi_1(X') \cong H$; thus the Albanese morphism $\mathrm{alb}_{X'} : X' \to \mathrm{Alb}(X')$ is a first approximation to $f(\sigma)$. G acts on $\mathrm{Alb}(X')$, and $U(\Pi) = \mathrm{Alb}(X')/G$ is our best candidate. There is a natural morphism $f : X \to U(\Pi)$.

Unfortunately, it can happen that $\mathrm{Alb}(X')/G$ is simply connected. For example, let A be an Abelian surface and S a K3 surface with a fixed-point-free involution τ. Let $X = A \times S/(-1, \tau)$. X is smooth, projective, and there is an exact sequence

$$0 \to \mathbb{Z}^4 \to \pi_1(X) \to \mathbb{Z}_2 \to 0.$$

$\mathbb{Z}_2$ acts by conjugation on $\mathbb{Z}^4$ via multiplication by -1. $X' = A \times S$ and the Albanese map is the projection onto A. The action of $\mathbb{Z}_2$ on A is multiplication by -1. $A/(-1)$ is called the *Kummer surface* of A and it is simply connected.

It is possible to define an orbifold structure on $A/(-1)$ that gives the expected orbifold fundamental group. This may be a general pattern, but in all cases that I know the orbifold is a quotient of a variety by a finite group action, so we do not get anything essentially new.

0.3.3 Shafarevich morphism. Let X be a proper variety with fundamental group Π and assume that $U(\Pi)$ and $f : X \to U(\Pi)$ both exist. One can take the Stein factorization

$$X \xrightarrow{sh_X} Sh(X) \longrightarrow U(\Pi).$$

The morphism sh_X should be viewed as the non-Abelian analog of the morphism a_X defined in (0.1.3). As before, it is possible to give an intrinsic characterization of sh_X that makes sense without assuming the existence of $U(\Pi)$:

Definition. Let X be a normal and proper variety. A normal and proper variety $Sh(X)$ and a morphism $sh_X : X \to Sh(X)$ are called the *Shafarevich variety* and the *Shafarevich morphism* of X if

(0.3.3.1) sh_X has connected fibers, and

(0.3.3.2) for every closed and connected subvariety $Z \subset X$

$$sh_X(Z) = \text{point} \quad \text{iff} \quad \text{im}[\pi_1(Z) \to \pi_1(X)] \quad \text{is finite.}$$

It is easy to see that $sh_X : X \to Sh(X)$ is unique if it exists.

Unfortunately, I am unable to prove that the Shafarevich morphism always exists. Its existence is connected with the Shafarevich conjecture about the universal covers of a smooth projective variety.

THEOREM. *Let X be a normal and proper variety. The following are equivalent:*

(0.3.3.3) $sh_X : X \to Sh(X)$ exists.

(0.3.3.4) The universal cover $\tilde{X}$ of X admits a proper morphism $sh_{\tilde{X}}$: $\tilde{X} \to Sh(\tilde{X})$ onto a complex space $Sh(\tilde{X})$ which does not have any positive dimensional compact complex subspaces.

In (3.5) a weaker version of this definition is considered where we require (0.3.3.2) only for "most" subvarieties $Z \subset X$. The corresponding map $\text{sh}_X : X \dashrightarrow \text{Sh}(X)$ is called the *Shafarevich map*. (See (0.4.7) for the distinction between the notions map and morphism.)

By (3.6) the Shafarevich map always exists.

0.3.4 Automorphic forms. Let X be a smooth projective variety and $u : M \to X$ the universal cover. The classical theory of automorphic forms assumes that M is well known, and uses this information to get results about the function theory of X.

In general M and X are both unknown, and we attempt to get information about them simultaneously by exploiting the existence of the universal covering morphism $u : M \to X$.

This seems a very interesting direction. The results of these notes are hopefully only the beginning steps in this area.

0.3.5 Varieties with generically large algebraic fundamental group. We would like to consider those varieties for which the Shafarevich map is birational. It is technically easier to study the following related property.

Definition. Let X be a proper variety. We say that X has *generically large algebraic fundamental group* if

$$\text{im}[\hat{\pi}_1(Z) \to \hat{\pi}_1(X)] \text{ is infinite for "most" subvarieties } Z \subset X.$$

Here $\hat{\pi}_1$ denotes the algebraic fundamental group, which is essentially the set of all finite quotients of π_1. The precise definitions are given in (4.2, 4.5).

Several of the results mentioned in (0.2.5) have their non-Abelian counterparts. For the proof and for further results, consult (16.3).

THEOREM. *Let X be a smooth projective variety of general type and assume that X has generically large algebraic fundamental group. Then* $H^0(X, K_X^2) \geq 1$.

CONJECTURE. *Let X be a smooth projective variety and assume that X has generically large fundamental group. Then* $\chi(X, K_X) \geq 0$.

0.3.6 Applications to pluricanonical maps. My main result in this direction is the following analog of (0.1.6):

THEOREM. [Kollár93b, 1.15] *Let X be a smooth projective threefold of general type such that $\hat{\pi}_1(X)$ is infinite. Then*
(0.3.6.1) $H^0(X, K_X^2) \geq 1$, *and*
(0.3.6.2) $H^0(X, K_X^k)$ *gives a birational map for* $k \geq 49$.

Description of the Chapters

Shafarevich maps are studied in the first four chapters (Part I). They were introduced in [Kollár93b] in order to find a substitute for the Shafarevich conjecture about the universal covers of algebraic varieties [Shafarevich72, IX.4.3]. Shafarevich maps were independently discovered by [Campana94] in the context of Kähler manifolds. (Some of these aspects are discussed in chapter 18.)

Chapter 1 presents an alternative motivation to Shafarevich maps by investigating the following Lefschetz-type problem.

Question. Let $Z \subset X$ be a subvariety of an algebraic variety. Assume that $\pi_1(Z) \to \pi_1(X)$ is not surjective. Is there some clear geometric reason for this?

There seems to be very little that one can say for arbitrary subvarieties, but for sufficiently general subvarieties a very nice picture emerges (1.6). Shafarevich maps are the key to understanding these questions.

Chapter 2 contains technical results needed in the proof of the existence of Shafarevich maps. Most of the necessary results are part of the basic theory of algebraic cycles and were known already in the thirties.

The proof of the existence of Shafarevich maps is presented in chapter 3. Some other variants are also considered. From the technical point of view the most useful is the *algebraic Shafarevich map*, which is defined in (4.3).

Chapter 4 considers the relationship between the fundamental group and the Kodaira dimension of a variety. The main implication of the results and conjectures is that varieties of general type are the only ones with an interesting fundamental group. The hardest (and most interesting) open problem is to prove that if X is a smooth projective variety with Kodaira dimension zero, then $\pi_1(X)$ has a finite index Abelian subgroup.

The next four chapters, in Part II, consider various aspects of the classical theory of automorphic forms. Most of these results are not necessary for the rest of the arguments. They, however, serve as guideposts in investigating the more general situation.

Chapter 5 contains a discussion about Poincaré series that works for bounded domains in Stein manifolds. I follow the presentation of [Siegel73].

Chapter 6 outlines some of the applications of the L^2 index theorem to universal covers of projective varieties. These results apply to any variety X, but do not readily yield information about automorphic forms without some vanishing theorems. This method is used repeatedly later to construct automorphic forms.

Chapter 7 contains the proof that for quotients of bounded symmetric domains every automorphic form of sufficiently high weight can be represented by Poincaré series. (The weight restriction turns out to be necessary as well.) My original hope was to extend the proof to other cases, at least to bounded nonsymmetric domains. Unfortunately, the method requires rather delicate information about the Bergman kernel that is not generally known. Nonsymmetric domains with compact quotients tend to have rather complicated boundaries. Because of the existence of a compact quotient, the Bergman kernel has strong periodicity properties, and hopefully this compensates for the loss of regularity of the boundary. It would be very interesting to see more results in this direction.

The example of ball quotients is discussed in chapter 8. The theory of fundamental groups of varieties suffers from a lack of examples. There are few cases where one can successfully compute both on X and on its universal cover. Ball quotients provide a reasonably large and interesting class of examples where such computations are possible. All these results can be generalized to bounded symmetric domains.

Chapters 9–12 of Part III about vanishing theorems form a side direction, which is, however, indispensable for the main results. They explore various generalizations of the Kodaira vanishing theorem developed by [Demailly82; Kawamata82; Viehweg82; Kollár86a; Esnault-Viehweg86; Nadel90b]. These theorems are not new. The lectures [Esnault-Viehweg92] discuss most of these results using the theory of De Rham complexes.

Chapter 9 contains a simpler topological argument in the spirit of [Kollár86b, sec. 5]. This is then generalized to the more refined vanishing and injectivity theorems in chapter 10.

The L^2-vanishing theorems of Demailly and Nadel are discussed in chapter 11, mostly without proofs. These are essentially equivalent to the algebraic versions.

Chapter 12 is even more of an aside. It discusses vanishing results on singular varieties. The concept of Du Bois singularities—introduced by [Steenbrink83]—emerges as the right definition.

The main results of these notes are presented in chapters 13 to 16 of Part IV.

Chapter 13 reviews the results of [Gromov91] generalizing the constructions of Poincaré outlined in chapter 5. The main advantage of this method is that it can be used to produce lots of automorphic forms using very little information about the universal cover.

The first main existence result on automorphic forms is proved in (14.5):

THEOREM. *Let X be a smooth and proper variety. Pick a sufficiently general point $x \in X$. Assume that if $x \in Z \subset X$ is an irreducible, positive dimensional subvariety, then $\hat{\pi}_1(Z')$ is infinite where $Z' \to Z$ is a resolution of singularities. Let L be a Cartier divisor on X such that sections of L^m separate points of a dense open set $X^0 \subset X$ for $m \gg 1$.*

Then $h^0(X, \mathcal{O}(K_X + L)) > 0$.

This is a rather curious result on two counts. First, it is not clear to me that the assumption about the fundamental groups of subvarieties is relevant. There are counterexamples without this assumption, but my feeling is that there should be more general theorems that assume much less about the subvarieties of X. Second, I do not see how one can use this method to get two independent sections of K_X^m, even for large values of m.

A method that produces at least two independent sections of K_X^m is explained in chapter 15. The main result is the multiplicativity of plurigenera of varieties of general type under étale covers. Comparing the plurigenera of X with the plurigenera of an étale cover, one can show that some plurigenus of X is at least two. Again we encounter the situation that the result fails without the assumptions about the fundamental group, but it is hard to believe that the fundamental group plays an essential role.

The existence theorems about automorphic forms are summarized in chapter 16. Some of these results are close to being optimal; others need considerable improvement. This applies especially to the surjectivity of the Poincaré series map (16.5) where the bounds given are very large.

My original hope was to reach a point where the general assertions say something new and interesting about automorphic forms on bounded symmetric domains. I still hope that some time in the future this does happen. For the moment, however, this seems far away.

In Part V, chapter 17 examines some applications of the above ideas to Abelian varieties. Abelian varieties are the simplest higher-dimensional varieties with nontrivial fundamental group, and many of the methods work very efficiently.

Chapter 18 is a collection of open problems and speculations.

Terminology

0.4.1. I use the words *line bundle* and *invertible sheaf* interchangeably. If D is a *Cartier divisor* on a variety X, then $\mathcal{O}_X(D)$ denotes the corresponding line bundle. *Linear equivalence* of line bundles (resp. Cartier divisors) is denoted by $\cong$ (resp. $\sim$). *Numerical equivalence* is denoted by $\equiv$.

0.4.2. A line bundle L on a variety X is called *big* if there is a dense open set $U \subset X$ such that if $x_1, x_2 \in U$ and $m \gg 1$ then there is a section $f \in H^0(X, L^m)$ such that $f(x_1) = 0$ and $f(x_2) \neq 0$.

The so-called Kodaira lemma asserts that L is big if for every line bundle M there is an $m > 0$ and an effective divisor E such that $L^m \cong M \otimes \mathcal{O}_X(E)$.

A line bundle L on a variety X is called *nef* if $\deg_C L \geq 0$ for every compact irreducible curve $C \subset X$.

A Cartier divisor D is called big (resp. nef) if the corresponding line bundle $L = \mathcal{O}_X(D)$ is big (resp. nef).

0.4.3. By a $\mathbb{Q}$-*divisor* on a variety X we mean a $\mathbb{Q}$-linear combination $D = \sum d_i D_i$, where the D_i are divisors. D is called $\mathbb{Q}$-Cartier if $\sum(md_i)D_i$ is an integral Cartier divisor for some $m > 0$.

The notions of big and nef extend to $\mathbb{Q}$-divisors by linearity.

0.4.4. The *canonical line bundle* of a smooth variety X is denoted by K_X. By definition, $K_X = (\det T_X)^{-1}$ where T_X is the holomorphic tangent bundle. Thus $c_1(K_X) = -c_1(X)$.

If X is a normal variety, there is a unique divisor class K_X on X such that

$$\mathcal{O}_{X-\mathrm{Sing}(X)}(K_X|X - \mathrm{Sing}(X)) \cong K_{X-\mathrm{Sing}(X)}.$$

K_X is called the *canonical class of* X. The switching between the divisor and line bundle versions should not cause any problems.

0.4.5. A *pluricanonical line bundle* (resp. a *pluricanonical divisor*) is any positive power K_X^m (resp. any positive multiple mK_X) of the canonical line bundle (resp. divisor). $h^0(X, K_X^m)$ is called the mth *plurigenus* of X. It is denoted by $P_m(X)$. The map given by sections of K_X^m (equivalently by the linear system $|mK_X|$) is called a *pluricanonical map* (or *m-canonical map*).

0.4.6. Let X be a variety over $\mathbb{C}$ (or a topological space). Its *fundamental group* is denoted by $\pi_1(X)$. Its *algebraic fundamental group* (cf. (4.2)) is denoted by $\hat{\pi}_1(X)$.

0.4.7. A *morphism* between algebraic varieties is assumed to be everywhere defined. It is denoted by a solid arrow $\to$. A *map* is defined only on a dense open set. It is sometimes called a rational or meromorphic map to emphasize this fact. It is denoted by a broken arrow $\dashrightarrow$.

0.4.8. A morphism $f : X \to Y$ (of varieties over $\mathbb{C}$) is called *étale* if it is a complex analytic local isomorphism.

Part I

Shafarevich Maps

CHAPTER 1

Lefschetz-Type Theorems for π_1

Let X be a smooth projective variety and $H \subset X$ a smooth hyperplane section (under some projective embedding $X \subset \mathbb{P}$). Lefschetz theory asserts that the topology of X and the topology of H are closely related. One of the simplest results is the following.

1.1 THEOREM. [Lefschetz24] *Let X be a smooth projective variety and $H \subset X$ a smooth hyperplane section. Then $\pi_1(H) \to \pi_1(X)$ is an isomorphism if* $\dim H \geq 2$ *and a surjection if* $\dim H = 1$.

We try to extend this result to arbitrary subvarieties $Z \subset X$ by connecting it to the following problem.

1.2 CONJECTURE. [Shafarevich72] *Let X be a smooth projective variety and $u : \tilde{X} \to X$ the universal cover. Then there is a proper holomorphic morphism (with connected fibers) $sh_{\tilde{X}} : \tilde{X} \to Sh(\tilde{X})$ onto a normal Stein space $Sh(\tilde{X})$.*

1.3 Definition. Assume that the Shafarevich conjecture is true for X. $\pi_1(X)$ acts on $\tilde{X}$ by deck transformations and $sh_{\tilde{X}}$ is $\pi_1(X)$-equivariant. Set $Sh(X) = Sh(\tilde{X})/\pi_1(X)$ and let

(1.3.1) $\quad sh_X : X \to Sh(X)$

be the corresponding morphism. $Sh(X)$ is called the *Shafarevich variety* of X and sh_X the *Shafarevich morphism.*

In order to see the connection with (1.1) we need to understand the fibers of the Shafarevich map sh_X. Let $Z \subset X$ be an irreducible subvariety such that $\mathrm{im}[\pi_1(Z) \to \pi_1(X)]$ is finite. Then its preimage in the universal cover $u^{-1}(Z) \subset \tilde{X}$ is the union of finite covers of Z and each of these is projective. $Sh(\tilde{X})$ is Stein, thus it has no positive dimensional proper complex subspaces. Therefore every irreducible component of $u^{-1}(Z)$ is contained in a fiber of $sh_{\tilde{X}}$ and Z is contained in a fiber of sh_X. Conversely, let $F \subset X$ be an irreducible component of a fiber of sh_X. $\mathrm{im}[\pi_1(F) \to \pi_1(X)]$ defines a covering of $F' \to F$. F' is contained in a fiber of $sh_{\tilde{X}}$, hence it is proper. Therefore $\operatorname{im}[\pi_1(F) \to \pi_1(X)]$ is finite.

Using the notation of (1.1) assume that $\operatorname{im}[\pi_1(H) \to \pi_1(X)]$ is finite. By the above considerations, then, H is contained in a fiber of sh_X and H is disjoint from the other fibers. Since H is ample, this is possible only if there is exactly one fiber, i.e., $Sh(X) =$ point. Thus $\pi_1(X)$ itself is finite. This is much weaker than (1.1) but at least the connection is becoming clearer.

Unfortunately, I have no idea how to prove the Shafarevich conjecture. The problem, however, becomes easier if we are interested only in "general subvarieties."

1.4 General subvarieties. The notion of "general" subvarieties should depend on the given situation. It turns out that in the present case a fairly simple definition suffices: we need to assume only that Z contains a "general" point of X. General points are easy to define.

1.4.1 Definition. Let **P** be a property thet makes sense for closed points of a given variety X. (For example, **P** can be "$x \in X$ is smooth," or "there is no rational curve in X containing x.") We say that **P** is satisfied by *general* (resp. *very general*) points of X if there are finitely many (resp. countably many) subvarieties $V_i \subsetneq X$ such that **P** holds for every point $x \in X$ outside $\cup V_i$.

I will frequently be sloppy and say instead that a *general* (resp. *very general*) point of X *satisfies* **P**.

1.4.2 Examples. (1.4.2.1) For any variety X, a general point of X is smooth.

(1.4.2.2) Let X be a smooth surface. If X is not birationally ruled, then there is no rational curve through a very general point of X.

The adjective "very" cannot be omitted from the above statement as shown by the following example.

Let $X \subset \mathbb{P}^3$ be a very general degree 4 surface. (Degree 4 surfaces are parametrized by a projective space of dimension 34, and the surface should correspond to a very general point in this space.) The argument of [Mori-Mukai83] shows that X contains countably many rational curves $C_i \subset X$.

The following examples show what one can expect about $\operatorname{im}[\pi_1(Z) \to \pi_1(X)]$ if X is an arbitrary quasi-projective variety and $Z \subset X$ is an arbitrary subvariety.

1.5 Examples. (1.5.1) Let us start with $H \subset X$ as in (1.1). Pick points $x_0, \ldots, x_k \in X - H$. It is not difficult to see that we can get a new projective variety X' by identifying the points $x_0, \ldots, x_k$. X' is not normal at the image of these points. It is easy to check that $\pi_1(X') = \pi_1(X) * F_k$, where F_k is the free group on k generators and $*$ denotes free product. Thus $\pi_1(H) \to \pi_1(X')$ is not surjective. This is the reason why normality is always assumed. (It is sufficient to assume that every point of X is analytically unibranch.)

(1.5.2) Let $0 \in \mathbb{P}^2$ be a fixed point and $L_1, \ldots, L_k$ lines through 0. Let $X = \mathbb{P}^2 - (L_1 \cup \ldots \cup L_k)$. X retracts to a line minus k points, thus $\pi_1(X) = F_{k-1}$. Let $L \cap X \subset X$ be the intersection of X with a line L. If L does not pass through 0, then $\pi_1(L \cap X) \to \pi_1(X)$ is an isomorphism. If L passes through 0, then $\pi_1(L \cap X) = \{1\}$, thus surjectivity fails.

(1.5.3) Let $H \subset \mathbb{P}^n$ be a closed subvariety and $X \subset \mathbb{P}^{n+1}$ the cone over H with vertex p. Let $B_pX \subset X \times H$ be the graph of the projection from p to H. $B_pX \to X$ is an isomorphism over $X - \{p\}$ and the fiber over p is a divisor E which is isomorphic to H. $B_pX \to H$ is a $\mathbb{P}^1$-bundle, thus $\pi_1(B_pX) \cong \pi_1(H)$. When we pass to X, we contract E, which carries all the fundamental group of B_pX. Thus $\pi_1(X) = \{1\}$. Therefore surjectivity is the most we can hope for in (1.1) if we allow singular varieties.

(1.5.4) Let X be a smooth variety of dimension at least 3 and $p : Y \to X$ a finite étale cover. Fix any point $x \in X$ and choose $y \in Y$ such that $p(y) = x$. Let $C \subset Y$ be a smooth curve containing y obtained as the intersection of general hypersurfaces. Since $\dim Y \geq 3$, $p : C \to p(C)$ is an isomorphism. $\pi_1(p(C)) \to \pi_1(X)$ factors as

$$\pi_1(p(C)) \xrightarrow{\cong} \pi_1(C) \to \pi_1(Y) \to \pi_1(X),$$

hence it is not surjective. Observe, however, that the image of $\pi_1(p(C)) \to \pi_1(X)$ has finite index in $\pi_1(X)$.

(1.5.5) Assume that there is a morphism $g : X \to Y$. Given any point $x \in X$ let $Z_x \subset X$ be the fiber of g containing x. If $\pi_1(Y)$ is infinite then $\pi_1(Z_x) \to \pi_1(X)$ is not surjective and the image does not have finite index in $\pi_1(X)$.

A more sophisticated version of this example is the following.

(1.5.6) Let Y be a proper variety that has an order 4 automorphism τ such that τ^2 is fixed point free. Let τ' be the automorphism of $Y \times Y$ given by $\tau'(y_1, y_2) = (\tau y_2, \tau y_1)$. Set $X = Y \times Y/\langle \tau' \rangle$. There is an exact sequence

(1.5.6.1) $\quad 1 \to \pi_1(Y) \times \pi_1(Y) \to \pi_1(X) \to \mathbb{Z}_4 \to 0.$

Pick any y_2 and consider the morphism $f : Y \to X$ given by $f(y) = (y, y_2) \bmod \langle \tau' \rangle$. f is an immersion and the image has exactly one double point since $f(\tau y_2) = f(\tau^3 y_2)$.

1.5.6.2 CLAIM. *Let $h : X \dashrightarrow Z$ be a map that is a morphism along $f(Y)$ and such that $h(f(Y))$ is a point. Then $h(X)$ is a point.*

Proof. By composition we obtain $h' : Y \times Y \dashrightarrow Z$. The preimage of $f(Y)$ in $Y \times Y$ is a union of four irreducible varieties:

$$Y \times \{y_2\} \cup Y \times \{\tau^2 y_2\} \cup \{\tau y_2\} \times Y \cup \{\tau^3 y_2\} \times Y.$$

By assumption, h' is defined along these and each is contracted to a point. Thus every horizontal and every vertical section of $Y \times Y$ is contracted to a point by h' (cf., e.g., [CKM88, 1.5]). Therefore $h'(Y \times Y)$ is a point. ∎

The image of $f_* : \pi_1(Y) \to \pi_1(X)$ is the first factor in (1.5.6.1), hence it has infinite index in $\pi_1(X)$ if $\pi_1(Y)$ is infinite. By (1.5.6.2) if a morphism contracts $f(Y)$ to a point, then it contracts X to the same point.

If we take the degree 4 étale cover $Y \times Y \to X$ and lift $f : Y \to X$ to $f' : Y \to Y \times Y$, then a projection map of $Y \times Y$ shows that the situation becomes entirely analogous to (1.5.5).

The above examples in fact exhaust the list of necessary exceptions. We can even be slightly more general and consider arbitrary morphisms $f : Y \to X$ instead of just normalizations of subvarieties.

1.6 THEOREM. *Let X and Y be normal, irreducible and quasi-projective varieties. Let $f : Y \to X$ be a morphism such that $f(Y)$ contains a very general point of X. Fix a point $y \in Y$. Then one can construct the following:*

(1.6.1.1) a finite étale cover $p : X' \to X$;
(1.6.1.2) a lifting of f to $f' : Y \to X'$ such that $f = p \circ f'$;
(1.6.1.3) an open set $U \subset X'$; and
(1.6.1.4) a morphism $g : U \to V$ onto a quasi-projective variety V;

such that the following properties are satisfied (set $x' = f'(y')$):

(1.6.2.1) every fiber of g is closed in X';
(1.6.2.2) g is a topologically locally trivial fiber bundle;
(1.6.2.3) $f(Y)$ is contained in a fiber U_v of g;
(1.6.2.4) $\operatorname{im}[\pi_1(Y, y) \to \pi_1(X', x')]$ *has finite index in* $\operatorname{im}[\pi_1(U_v, x') \to \pi_1(X', x')]$.

1.6.3 Remark. An a priori definition of "very general" is given in (2.6).

1.6.4 Comment. In (1.6) we assume that x is a very general point of X. Thus we exclude a "bad" subset, which is a countable union of subvarieties. This may not be necessary. In all the examples I know, a finite union would be sufficient.

1.7 Example. In (1.6.2.4) it would be natural to hope that $\operatorname{im}[\pi_1(Y, y) \to \pi_1(U_v, f'(y))]$ has finite index in $\pi_1(U_v, f'(y))$. Unfortunately, this is not always possible to achieve.

Let $\mathbb{C}^2$ be a plane with coordinates (x_1, x_2) and C_i smooth projective models of the hyperelliptic curves $y_i^2 = f(z_i)$. Consider the action of the dihedral group D_4 on $\mathbb{C}^2$ and on $C_1 \times C_2$ generated by

$$\tau_1(x_1, x_2) = (-x_1, x_2), \qquad \tau_1(y_1, y_2, z_1, z_2) = (-y_1, y_2, z_1, z_2),$$
$$\tau_2(x_1, x_2) = (x_1, -x_2), \qquad \tau_2(y_1, y_2, z_1, z_2) = (y_1, -y_2, z_1, z_2),$$
$$\sigma(x_1, x_2) = (x_2, x_1), \qquad \sigma(y_1, y_2, z_1, z_2) = (y_2, y_1, z_2, z_1).$$

This gives a faithful action on the conic $Q = (x_1^2 + x_2^2 = x_0^2) \subset \mathbb{P}^2$.

Let $Z = Q \times C_1 \times C_2/D_4$ where we act by the above formulas. Let $q : Z \to Q/D_4 \cong \mathbb{P}^1$ be the first coordinate projection and $B \subset \mathbb{P}^1$ the set of singular fibers. The general fiber of q is isomorphic to $C_1 \times C_2$ and we have a monodromy representation $\pi_1(\mathbb{P}^1 - B) \to \operatorname{Aut}(C_1 \times C_2)$ whose image is D_4.

Let $p \in \mathbb{P}^1$ be a fixed point of τ_1. τ_1 acts on $\{p\} \times C_1 \times C_2$ and the quotient is isomorphic to $\mathbb{P}^1 \times C_2$. Thus the map $\pi_1(C_1) \to \pi_1(Z)$ factors through $\pi_1(\mathbb{P}^1) = 1$. Hence $\operatorname{im}[\pi_1(C_1) \to \pi_1(Z)] = \{1\}$ and similarly for C_2. This shows that Z is simply connected. (Z has some quotient singularities; these can be resolved without changing the fundamental group [Kollár93b, 7.8].)

Let D be any curve of positive genus and $h : D \to \mathbb{P}^1$ a morphism that is unramified over B. Let $X = Z \times_{\mathbb{P}^1} D$. Then $\pi_1(X) \cong \pi_1(D)$. As before, the general fiber of $X \to D$ is isomorphic to $C_1 \times C_2$ and $\operatorname{im}[\pi_1(C_1) \to \pi_1(X)] = \{1\}$. Also, the image of the monodromy representation $\pi_1(D - h^{-1}(B)) \to \operatorname{Aut}(C_1 \times C_2)$ is still D_4.

Set $Y = C_1$ and let $f : Y \to X$ be the embedding of C_1 as one of the factors of a smooth fiber of $X \to D$. If $g : X \dashrightarrow V$ is a morphism with connected fibers such that g is defined along $f(Y)$ and contracts it to a point then the monodromy forces it to contract the whole fiber containing $f(Y)$. $\operatorname{im}[\pi_1(C_1) \to \pi_1(C_1 \times C_2)]$ has infinite index in $\pi_1(C_1 \times C_2)$.

If $D' \to D$ is a finite étale cover of X, then $h' : D' \to D \to \mathbb{P}^1$ is still unramified over B. Thus even after taking étale covers of X we cannot have a morphism as in (1.6) such that $\operatorname{im}[\pi_1(Y, y) \to \pi_1(U_v, f'(y))]$ has finite index in $\pi_1(U_v, f'(y))$.

It is nonetheless possible to control the fundamental group of U_v in terms of the fundamental group of Y. The full result is stated in (3.11).

Here we mention an important special case where the formulation is very simple.

1.8 THEOREM. *Let X be a proper and normal variety. There is an open set $X^0 \subset X$ and a proper morphism $f^0 : X^0 \to Z^0$ with the following properties:*

(1.8.1) f^0 is a topologically locally trivial fibration.

(1.8.2) For every $z \in Z^0$ the fundamental group of X_z^0 is finite.

(1.8.3) If $z \in Z^0$ is very general and $w : W \to X$ is a morphism from an irreducible normal variety W such that $\pi_1(W)$ is finite and $\operatorname{im} w$ *intersects X_z^0, then* $\operatorname{im} w \subset X_z^0$.

The proofs of (1.6) and (1.8) are given in chapter 3. Here we concentrate instead on explaining the result and pointing out some corollaries.

We would like to assert that in general $\operatorname{im}[\pi_1(Y) \to \pi_1(X)]$ is "large" in $\pi_1(X)$. This is indeed the case if for some reason there are very few morphisms $U \to V$ satisfying the conditions (1.6). To get some nice examples assume for simplicity that X is proper. In this case every fiber of g is proper, thus g itself is proper.

1.9 LEMMA. *Let X be a smooth projective variety. Assume that* $\operatorname{rank}\operatorname{Pic}(X) = 1$. *Let $U \subset X$ be an open subset and $g : U \to V$ a proper morphism with positive dimensional fibers. Then $X = U$ and $V =$ point.*

Proof. Assume that $\dim V > 0$. Let $H_V \subset V$ be an effective divisor and $H_X \subset X$ the closure of $g^{-1}(H_V)$. Then H_X is disjoint from the general fiber of g. By assumption, every effective divisor on X is ample, and therefore it intersects every positive dimensional subvariety. Thus g has zero dimensional fibers. ∎

Combining (1.8) and (1.9) we obtain the following.

1.10 COROLLARY. *Let X be a smooth projective variety such that $\pi_1(X)$ is infinite. Assume that* $\operatorname{rank}\operatorname{Pic}(X) = 1$. *Let $Z \subset X$ be an irreducible positive dimensional subvariety containing a very general point. Then* $\operatorname{im}[\pi_1(\bar{Z}) \to \pi_1(X)]$ *is infinite.* ∎

1.10.1 Example. The following variant of (1.5.6) shows that in general $\operatorname{im}[\pi_1(\bar{Z}) \to \pi_1(X)]$ does not have finite index in $\pi_1(X)$.

Let H be the moduli space of principally polarized Abelian varieties of dimension 8 with level 3 structure (cf. [Mumford-Fogarty82, p. 129]). H is a smooth quasi-projective variety whose fundamental group Γ is the level 3 congruence subgroup of $Sp(8, \mathbb{Z})$. $Sp(8, \mathbb{Z})/\Gamma$ acts on H. Let $\rho : \mathbb{Z}_4 \to GL(4, \mathbb{Z})$ be the regular representation of $\mathbb{Z}_4$. We obtain a block diagonal representation

$$\tau : \mathbb{Z}_4 \to Sp(8, \mathbb{Z}) \quad \text{by } \tau(m) = \operatorname{diag}(\rho(m), \rho(m), \rho(m), \rho(m)).$$

It is easy to see that the fixed point set of $\tau(2)$ has codimension 12 in H. Let $Y \subset H$ be a general $\tau(1)$-invariant threefold section of H.

Using this Y in (1.5.6) we obtain $X = (Y \times Y)/\langle \tau' \rangle$. Let $Z \subset X$ be the image of $Y \times \{y\}$ for any $y \in Y$. Then $\mathrm{im}[\pi_1(\bar{Z}) \to \pi_1(X)]$ does not have finite index in $\pi_1(X)$ by (1.5.6) and $\operatorname{rank} \operatorname{Pic} X = 1$.

The Shafarevich conjecture implies that (1.10) holds for arbitrary subvarieties.

1.10.2 CONJECTURE. *Let X be a smooth projective variety such that $\pi_1(X)$ is infinite. Assume that* $\operatorname{rank} \operatorname{Pic}(X) = 1$. *Let $Z \subset X$ be an irreducible positive dimensional subvariety. Then* $\mathrm{im}[\pi_1(\bar{Z}) \to \pi_1(X)]$ *is infinite.*

In particular, X does not contain any rational curve.

This is not known, even for smooth surfaces. [Nori83] treated the case when $\dim X = 2$ and $Z \subset X$ is a nodal curve with "not too many" nodes.

Another case where one can completely describe the possible morphisms $g : U \to V$ is the following.

1.11 Example. Let A be an Abelian variety. Let $U \subset A$ be an open subset and $g : U \to V$ a proper morphism with connected fibers. Then there is an Abelian subvariety $B \subset A$ such that $g : U \to V$ is birational to the quotient morphism $A \to A/B$.

Proof. We may assume that V is affine. Let U_v be a general fiber of g. Then $U_v \subset X$ is a closed and smooth subvariety. Let $a \in A$ be sufficiently close to the origin 0, which we assume to be in U_v. Then $a(U_v) \subset U$ and $g(a(U_v)) \subset V$. Since V is affine and U_v is proper, this implies that $g(a(U_v))$ is a point for every a. Thus the fibers of g are translates of U_v. Furthermore, if $a(U_v)$ intersects U_v, then $a(U_v) = U_v$. Let $b \in U_v$ close to 0. Then $b \in b(U_v) \cap U_v$, hence $b(U_v) = U_v$. Thus U_v equals the algebraic subgroup generated by the points of U_v, hence U_v is an Abelian subvariety. ∎

If $Z \subset X$ is a divisor, then $\pi_1(\bar{Z}) \to \pi_1(X)$ is almost always large.

1.12 THEOREM. *Let X be a normal and projective variety. Let $|H|$ be a (not necessarily complete) linear system with a base point $x \in X$. There are only finitely many irreducible divisors $H_t \in |H|$ such that* $\mathrm{im}[\pi_1(\bar{H}_t) \to \pi_1(X)]$ *does not have finite index in $\pi_1(X)$.*

This is proved in (2.12). Even in very nice situations as in (1.12), surjectivity usually fails.

1.13 Examples. (1.13.1) Let X be a smooth projective surface with a finite étale cover $p : Y \to X$. Let H_X be ample on X such that p^*H_X is very ample in Y. Let $C_Y \in |p^*H_X|$ be a general member passing through a point of $p^{-1}(x)$ for some fixed $x \in X$. Then $C_Y \to p(C_Y)$ is birational and $p(C_Y) \in |\deg p \cdot H_X|$. $\pi_1(\overline{p(C_Y)}) \to \pi_1(X)$ factors through $\pi_1(Y)$, thus it is not surjective.

(1.13.2) Let S be a K3 surface with a fixed-point-free involution τ and C a hyperelliptic curve with involution σ. Pick $s \in S$ and $c \in C$ such that c is not a fixed point of σ. Let $C_1 = \{s\} \times C$, $C_2 = \{\tau s\} \times C$. Let $p : W' \to S \times C$ be obtained from $S \times C$ by blowing up the four points $(s, c), (\tau s, c), (s, \sigma c), (\tau s, \sigma c)$. Let $E \subset W'$ be the (reducible) exceptional divisor.

Let M be a sufficiently ample divisor on S and $L = \mathcal{O}_C(2)$ σ-invariant. It is easy to see that $p^*(\pi_S^* M \otimes \pi_C^* L)(-E)$ is base-point-free and contracts the birational transforms of the curves C_i to points. By Stein factorization we obtain a threefold W with two singular points. The involution $\rho(u, v) = (\tau u, \sigma v)$ descends to a fixed-point free involution on W. Let X be the quotient with the unique singular point $x \in X$. The curves C_i carry all the fundamental group of $S \times C$, thus $\pi_1(X) = \mathbb{Z}_2$. The projection $\pi_C : S \times C \to C$ descends to a map $h : X \dashrightarrow C/\sigma \cong \mathbb{P}^1$. h gives a pencil $|H|$ on X with a base point at x. The general member of H is the image of a K3 surface in X with two points pinched together at x. Thus $\pi_1(\bar{H}_t) \to \pi_1(X)$ is not surjective.

(1.13.3) In (1.5.6) take Y to be a curve of genus at least 1. The image of $Y \times \{y_0\}$ is a divisor on X, which shows that the finitely many exceptions in (1.12) are indeed necessary.

CHAPTER 2

Families of Algebraic Cycles

In this chapter I will set up the machinery of families of algebraic cycles on a variety X. Since we deal with properties of X at very general points only, we are able to ignore all the subtle aspects of cycle theory. Almost everything that we need about cycles is contained in [Hodge-Pedoe52].

2.1 Definition. Let X be a normal variety. By a *normal cycle* on X we mean an irreducible and normal variety W together with a finite morphism $w : W \to X$, which is birational to its image.

Let $Z \subset X$ be any closed irreducible subvariety, and $n : \bar{Z} \to Z \subset X$ the normalization. This is a normal cycle on X.

2.2 Definition. Let X be a normal variety. A *family of normal cycles* on X is a diagram

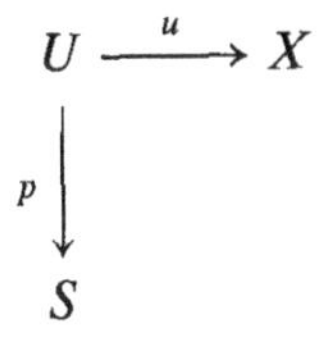

where

(2.2.1) every connected component of U and of S is of finite type and reduced (but there can be countably many such components);

(2.2.2) p is flat with irreducible, geometrically reduced and normal fibers;

(2.2.3) for every $s \in S$, $u|U_s : U_s \to X$ is a normal cycle. (Here and later U_s stands for the fiber of p over $s \in S$.)

We usually write only part of the above diagram (the most frequent versions are $u : U \to X$ or $p : U \to S$) when the rest is clear from the context. The morphism u is called the *cycle map*.

2.3 Definition. (2.3.1) We say that the family $u : U \to X$ is *dominant* if u is dominant. We usually use this notion only if S is irreducible.

(2.3.2) We say that $p : U \to S$ is a *locally topologically trivial family* of normal cycles if $U \to S$ is a topological fiber bundle.

We are finally able to give a precise meaning to "very general" in (1.6).

2.4 Definition. Let X be a normal variety. Let $VG(X) \subset X$ be the collection of those closed points $x \in X$ that satisfy the following property.

For every normal cycle $w_0 : W_0 \to X$ such that $x \in \operatorname{im} w_0$ there is a *dominant* and locally topologically trivial family of normal cycles $S \longleftarrow W \xrightarrow{w} X$ and a closed point $s \in S$ such that $[w : W_s \to X] \cong [w_0 : W_0 \to X]$.

We see in (2.5) that $VG(X)$ is not empty. $VG(X)$ seems very difficult to compute, except in some simple examples.

2.4.1 Examples. (2.4.1.1) Assume that an algebraic group acts on X with a dense orbit $U \subset X$. Then $VG(X) \supset U$.

(2.4.1.2) Let X be a smooth projective surface and $f_0 : C_0 \to X$ a normal cycle, i.e., f_0 is a morphism from a smooth curve to X. Let C_t be a flat deformation of C_0 and assume that f_0 can be extended to a family of morphisms $f_t : C_t \to X$. Then $f_0 : C_0 \to X$ is part of a dominant locally topologically trivial family of normal cycles, unless $C_t \cong C_0$ and f_t is the composition of f_0 with an automorphism of C_0 for every value of t.

It is not too difficult to see that the nontrivial deformations of the pair (f_0, C_0) form a family of dimension at least

$$\deg f_0^* T_X + 2(1-g) + \dim \mathcal{M}_g - \dim \operatorname{Aut}(C_0),$$

where $g = g(C_0)$ and $\mathcal{M}_g$ is the moduli space of genus g curves. If $-K_X$ is nef, then we obtain nontrivial deformations unless C_0 is rational and $-K_X \cdot C_0 = 0, 1$, or C_0, is elliptic and $-K_X \cdot C_0 = 0$. This way we obtain

(2.4.1.2.1) If X is a Del Pezzo surface, then $VG(X) = X - \{(-1)\text{-curves}\}$.

(2.4.1.2.2) Let X be obtained from $\mathbb{P}^2$ by blowing up 9 general points. Let $E \subset X$ be the birational transform of the unique cubic containing all nine points. Then $VG(X) = X - E - \{(-1)\text{-curves}\}$. As was noted by Zariski, X contains infinitely many (-1)-curves.

(2.4.1.2.3) Let X be a K3 surface. By [Ran94, 5.1], if E is an elliptic curve and $f : E \to X$ a morphism, then the pair (f, E) can be deformed. This and the above estimate gives that

$$VG(X) = X - \{\text{rational curves}\}.$$

In some cases X contains infinitely many rational curves.

2.5 PROPOSITION. *Notation as above. There are countably many closed subvarieties $D_\iota \subset X$ $(D_\iota \neq X)$ such that $\cup D_\iota$ contains the complement of $VG(X)$.*

Proof. This follows from (2.9). ■

2.6 COMPLEMENT TO (1.6). *For the purposes of (1.6) any point of $VG(X)$ is very general.*

This is proved along with (1.6) in chapter 3.

2.7 *Definition.* Notation as above. We say that $U \to S$ is a *weakly complete* locally topologically trivial family of normal cycles on X if

(2.7.1) $U \to S$ is a locally topologically trivial family of normal cycles on X, and

(2.7.2) for every locally topologically trivial family of normal cycles $W \to T$ there is a dense open set $T^0 \subset T$ and a unique morphism $t^0 : T^0 \to S$ such that the pullback of $U \to S$ by t^0 is isomorphic to the restriction of $W \to T$ to T^0:

$$\begin{array}{ccccccc} W \times_T T^0 & \longrightarrow & X & & U \times_S T^0 & \longrightarrow & X \\ \downarrow & & & \cong & \downarrow & & \\ T^0 & & & & T^0 & & \end{array}$$

2.8 PROPOSITION. *Let X be a normal variety. There is a weakly complete locally topologically trivial family of normal cycles*

$$S(X) \xleftarrow{p(X)} U(X) \xrightarrow{u(X)} X.$$

$S(X)$ has countably many connected components. We can even assume that $S(X)$ is smooth.

The proof is a rather straightforward application of some constructibility properties of algebraic morphisms. As for most results of this kind, the proof is not very illuminating.

Proof. It is difficult to deal with cycles on open varieties. Our aim is to try to understand cycles on X in terms of cycles on a compactification. Choose a compactification $\bar{X} \supset X$ and let $Z = \bar{X} - X$. There is a one-to-one correspondence between normal cycles on X and those normal cycles on $\bar{X}$ which are not contained in Z.

If $T \longleftarrow W \xrightarrow{w} \bar{X}$ is a normal cycle, set $W_Z = w^{-1}(Z)$ (set theoretically). We call the pair (W, W_Z) a normal cycle on the stratified space $(\bar{X}, Z)$. If $u : U \to \bar{X}$ is a family of normal cycles, then we say that it is a locally topologically trivial family of stratified normal cycles if $s : (U, U_Z) \to S$ is a topological fiber bundle of stratified spaces. This in particular implies that $s : U - U_Z \to S$ is a topological fiber bundle.

The following lemma enables us to achieve topological triviality of certain families.

2.8.1 LEMMA. *Let $h : Y_1 \to Y_2$ be a proper morphism between varieties. Let $Z_1 \subset Y_1$ be a closed subvariety. Then there is a dense Zariski open set $Y_2^0 \subset Y_2$ such that*

$$h|h^{-1}(Y_2^0) : (h^{-1}(Y_2^0), h^{-1}(Y_2^0) \cap Z_1) \to Y_2^0$$

is a topological fiber bundle of stratified spaces.

Proof. Choose a Whitney stratification (see, e.g., [Goresky-MacPherson88, I.1.2]) of Y_1 such that Z_1 is the union of strata. Y_2 has an open and dense subset $Y_2^0 \subset Y_2$ such that every stratum is smooth over Y_2^0. By [Goresky-MacPherson88, I.1.11] $h^{-1}(Y_2^0) \to Y_2^0$ is a topological fiber bundle of stratified spaces. ∎

Returning to the proof of (2.8), start with a universal family r : $\mathrm{Univ}(\bar{X}) \to \mathrm{Hilb}(\bar{X})$ of all cycles or subschemes of $\bar{X}$, where $\mathrm{Hilb}(\bar{X})$ is the Hilbert scheme. (We could as well use $\mathrm{Chow}(\bar{X})$ instead of $\mathrm{Hilb}(\bar{X})$.) If $\bar{X}$ is projective, the existence of $\mathrm{Chow}(\bar{X})$ is classical [Hodge-Pedoe52, X.8] and the existence of $\mathrm{Hilb}(\bar{X})$ is in [Grothendieck62]. For arbitrary $\bar{X}$, $\mathrm{Hilb}(\bar{X})$ is constructed in [Artin69] and $\mathrm{Chow}(\bar{X})$ in [Barlet75].

Replace $\mathrm{Hilb}(\bar{X})$ by $\mathrm{red}\,\mathrm{Hilb}(\bar{X})$ (same point set, but eliminate the nilpotents) and take the largest open subset which parametrizes reduced and irreducible cycles or subschemes. We use R to denote the resulting scheme. It is clear that set theoretically R is the same as the required $S(X)$. We need to find a suitable stratification of R such that $S(X)$ is the scheme-theoretic union of all strata.

Let $\bar{r} : \overline{\mathrm{Univ}} \to R$ be the normalization. (It is important that we do not normalize R.) There is a dense open subset $R_0 \subset R$ such that $\bar{r}$ is flat over R_0 and every fiber of $\bar{r}$ over R_0 is the normalization of the corresponding fiber of r. We may also assume that R_0 is smooth. Restrict r (not $\bar{r}$!) to $R - R_0$ and iterate this procedure to obtain R_1, and so on. We end up with countably many families $r_i : V_i \to R_i$ such that every normal cycle appears exactly once. Let $q : V \to Q$ be the union of all these with cycle map $v : V \to \bar{X}$.

By (2.8.1) there is a dense open set $Q_0 \subset Q$ such that

$$q|q^{-1}(Q_0) : (q^{-1}(Q_0), q^{-1}(Q_0) \cap v^{-1}(Z)) \to Q_0$$

is a topological fiber bundle of stratified spaces. We may also assume that Q_0 is smooth. Applying (2.8.1) to the family over $Q - Q_0$, we obtain Q_1, and so on. At the end we obtain families $q_i : W_i \to Q_i$ with cycle maps $w_i : W_i \to \bar{X}$ such that every $(W_i, w_i^{-1}(Z)) \to Q_i$ is a topological fiber bundle of stratified spaces.

In general we have some irreducible components $Q_i^j \subset Q_i$ such that $W_i^j \subset w_i^{-1}(Z)$; these induce the empty cycle on X. Let $S(X)$ be the union of all the other irreducible components of Q_i for every i and $U(X)$ the union of all the corresponding $W_i^j - w_i^{-1}(Z)$. It is clear from the construction that $U(X) \to S(X)$ is a locally topologically trivial family of normal cycles on X. We view $S(X)$ as the union of strata of a stratification of $R \subset \operatorname{red} \operatorname{Hilb}(\bar{X})$.

We still need to see that $U(X) \to S(X)$ is weakly complete. Let $W \to T$ be a family of normal cycles on X; assume that T is irreducible. Let $\bar{W} \supset W$ be the unique reduced scheme such that the cycle map $W \to T \times X$ extends to a finite morphism $\bar{W} \to T \times \bar{X}$. By generic flatness, there is a dense, Zariski open $T' \subset T$ such that the corresponding $\bar{W}' \to T'$ is flat with irreducible and reduced fibers. This gives a morphism $T' \to R \subset \operatorname{Hilb}(\bar{X})$. $S(X)$ is just a stratification of R, so there exists a unique connected component $S_i \subset S(X)$ such that $S_i \cap \operatorname{im}[T' \to R]$ is dense in $\operatorname{im}[T' \to R]$. Let $T^0 \subset T'$ be the largest open set whose image is contained in S_i. The requirements of (2.7.2) are satisfied by the universal property of the Hilbert scheme. ∎

2.8.2 Remark. The constructed $S(X)$ is not unique but fortunately we do not need uniqueness beyond what is required in (2.7). It is not clear to me if the functor "locally topologically trivial families of normal cycles" is coarsely representable or not. It is quite possible that one needs to change the definition.

2.9 COROLLARY. *Let X be a normal variety. There are countably many closed subvarieties $D_i \subset X$ ($D_i \neq X$) such that if $w : W \to X$ is a normal cycle and $\operatorname{im} w \not\subset \cup D_i$ then there is a unique point $s \in S(X)$ such that*

(2.9.1) $[u(X) : U_s(X) \to X] \cong [w : W \to X]$, and

(2.9.2) if $S_j(X) \subset S(X)$ denotes the irreducible component containing s, then $u_j(X) : U_j(X) \to X$ is dominant and flat over $X - D_j$.

Proof. Let $S_i(X)$ be the irreducible components of $S(X)$. Let D_i be the closure of $u(X)(U_i(X))$ if $U_i(X) \to X$ is not dominant. If $U_i(X) \to$

X is dominant, then it is flat over an open and dense set $X_j \subset X$. Set $D_j = X - X_j$. ∎

Let us end this chapter by proving some results about fundamental groups of algebraic varieties.

2.10 PROPOSITION. *(2.10.1) (cf. [Fulton-Lazarsfeld81, 0.7.B]) Let Y be a connected normal complex space and $Z \subset Y$ a Zariski closed subspace. Then $\pi_1(Y - Z) \to \pi_1(Y)$ is surjective.*

(2.10.2) (cf. [Campana91, 1.3]) Let X, Y be irreducible algebraic varieties, Y normal. Let $f : X \to Y$ be a dominant morphism such that the geometric generic fiber has k irreducible components. Then the index

$$[\pi_1(Y) : \operatorname{im}[\pi_1(X) \to \pi_1(Y)]]$$

divides k.

Proof. Let $h : Y' \to Y$ be the universal cover and $Z' = h^{-1}(Z)$. We need to show that $Y' - Z'$ is connected. $Y' - Z'$ is dense in Y', thus it is sufficient to prove that it is locally connected. Y' is normal and Z' is Zariski closed. Thus if $y' \in Y'$ is an arbitrary point and $y' \subset U \subset Y'$ is a sufficiently nice contractible neighborhood, then $U \cap (Y' - Z')$ is connected (see, e.g., [Gunning-Rossi65, III.A.10]). This shows (2.10.1).

In order to prove (2.10.2) we can factor f through a proper morphism; thus we may assume that f itself is proper. Let $Y^0 \subset Y$ be open and $X^0 = f^{-1}(Y^0)$. By (2.10.1) $\pi_1(X^0) \to \pi_1(X)$ and $\pi_1(Y^0) \to \pi_1(Y)$ are surjective; thus it is sufficient to prove (2.10.2) for $f^0 : X^0 \to Y^0$.

Let $X^0 \to Z^0 \to Y^0$ be the Stein factorization. By further shrinking Y^0, we may assume that $Z^0 \to Y^0$ is finite and is étale of degree k.

We may also assume that $\pi_1(X^0) \to \pi_1(Z^0)$ is surjective. Indeed, by (2.8.1) for suitable choice of Y^0, $X^0 \to Z^0$ is a topological fiber bundle with connected fibers.

Thus $\pi_1(X^0) \to \pi_1(Z^0)$ is surjective, $\pi_1(Z^0) \to \pi_1(Y^0)$ is injective, and the image has index dividing k. ∎

(2.10) has several consequences that may seem rather surprising.

2.11 PROPOSITION. *Let $f : Y \to X$ be a morphism between irreducible, normal varieties. Assume that* $\operatorname{im} f \cap VG(X) \neq \emptyset$. *Then the normalizer of* $\operatorname{im}[\pi_1(Y) \to \pi_1(X)]$ *is a finite index subgroup of* $\pi_1(X)$.

Proof. Assume first that $f : Y \to X$ is a normal cycle on X. By assumption there is a dominant and locally topologically trivial family of normal cycles

$$S \xleftarrow{p} \mathbf{Y} \xrightarrow{F} X$$

extending f. By the exact homotopy sequence of a fiber bundle, $\operatorname{im}[\pi_1(Y) \to \pi_1(\mathbf{Y})]$ is a normal subgroup of $\pi_1(\mathbf{Y})$. Thus $\operatorname{im}[\pi_1(Y) \to \pi_1(X)]$ is a normal subgroup of $\operatorname{im}[\pi_1(\mathbf{Y}) \to \pi_1(X)]$. By (2.10) the latter has finite index in $\pi_1(X)$.

In order to see the general case, let $w : W \to X$ be the normalization of the closure of $\operatorname{im} f$. Let $N < \pi_1(X)$ be the normalizer of $G = \operatorname{im}[\pi_1(W) \to \pi_1(X)]$. By (2.10) $\operatorname{im}[\pi_1(Y) \to \pi_1(X)]$ is a finite index subgroup of G. N acts on G by conjugation, and $\operatorname{im}[\pi_1(Y) \to \pi_1(X)]$ has only finitely many conjugates by (2.11.2). Thus a finite index subgroup of N normalizes $\operatorname{im}[\pi_1(Y) \to \pi_1(X)]$. ∎

2.11.1 Example. There are many infinite groups G such that if $H < G$ is a finitely generated subgroup such that $N_G(H)$ has finite index in G, then either H is finite or it has finite index in G. Thus if $G = \pi_1(X)$ and $f : Y \to X$ is a normal cycle through a very general point, then $\operatorname{im}[\pi_1(Y) \to \pi_1(X)]$ is either finite or has finite index in $\pi_1(X)$.

Such are, for instance, free groups and arithmetic subgroups of real simple Lie groups of real rank at least 2 (see, e.g., [Margulis91, p. 3])

2.11.2 LEMMA. *Let G be a finitely generated group. For every $m < \infty$, G has only finitely many subgroups of index m.*

Proof. Let $H < G$ be a subgroup of index m. We obtain a permutation representation on G/H by left multiplication. Thus H is determined by a homomorphism $G \to S_m$ and by the choice of an element of $\{1, \dots, m\}$. ∎

The following theorem can be viewed as a semicontinuity result for the fundamental groups of fibers of morphisms of complex spaces.

2.12 THEOREM. *Let X and Y be irreducible normal complex spaces and $f : X \to Y$ a morphism. Assume that there is a Zariski open dense set $Y^0 \subset Y$ such that $f : X^0 := f^{-1}(Y^0) \to Y^0$ is a topological fiber bundle with connected typical fiber X_g. Let $y \in Y$ be a point such that there is an $x \in f^{-1}(y)$ satisfying $\dim_x f^{-1}(y) = \dim X - \dim Y$. Then*

(2.12.1) there is an open neighborhood $y \in U \subset Y$ such that $\operatorname{im}[\pi_1(X_g) \to \pi_1(f^{-1}(U))]$ has finite index in $\pi_1(f^{-1}(U))$;

(2.12.2) if f is proper, then $\operatorname{im}[\pi_1(X_g) \to \pi_1(f^{-1}(y))]$ has finite index in $\pi_1(f^{-1}(y))$;

(2.12.3) if f is smooth at x, then $\pi_1(X_g) \to \pi_1(f^{-1}(U))$ is surjective.

Proof. The maps are defined as follows. Pick a point $y' \in U \cap Y^0$. Then X_g is homeomorphic to $f^{-1}(y') \subset U$. This defines $\pi_1(X_g) \to \pi_1(f^{-1}(U))$ (up to conjugation). If f is proper, then $f^{-1}(y)$ is a deformation retract of $f^{-1}(U)$ for suitable U. By composition we obtain $\pi_1(X_g) \to \pi_1(f^{-1}(y))$.

Let $x \in V \subset X$ be the intersection of X with $\dim X - \dim Y$ general hyperplanes (in a local embedding). Then $f : V \to Y$ is finite and open above a neighborhood $x \in U \subset Y$. We may also assume that $V \cap f^{-1}(U)$ retracts to x. Let $U^0 = U \cap Y^0$ and $V^0 = V \cap f^{-1}(U^0)$.

$X \times_Y V^0 \to V^0$ is a topological fiber bundle with typical fiber X_g and a section. Thus there is a right split exact sequence

$$\pi_1(X_g) \to \pi_1(X \times_Y V^0) \leftrightarrows \pi_1(V^0) \to 1.$$

$X \times_Y V^0 \to f^{-1}(U_0)$ is a finite covering and $f^{-1}(U_0)$ is Zariski open in $f^{-1}(U)$. Thus by (2.10), $\operatorname{im}[\pi_1(X \times_Y V^0) \to \pi_1(f^{-1}(U))]$ has finite index in $\pi_1(f^{-1}(U))$.

$$\operatorname{im}[\pi_1(V^0) \to \pi_1(f^{-1}(U))] \subset \operatorname{im}[\pi_1(V) \to \pi_1(f^{-1}(U))] = 1.$$

This shows (2.12.1).

If f is proper, then $f^{-1}(y)$ is a deformation retract of $f^{-1}(U)$ for suitable choice of U. Thus (2.12.1) implies (2.12.2).

Finally, if f is smooth at x, then $V^0 \to U^0$ is an isomorphism; thus $\pi_1(X \times_Y V^0) \to \pi_1(f^{-1}(U))$ is surjective by (2.10.1). ∎

2.12.4 COROLLARY. *Let X be a complex manifold and $f : X \to \Delta$ a proper morphism to the unit disc. Let $X_0 := f^{-1}(0) = \sum d_i Z_i$ and X_g the general fiber, assumed to be connected. Set $r = \gcd(d_i)$. There is an exact sequence*

$$\pi_1(X_g) \to \pi_1(X_0) \to \mathbb{Z}_r \to 0.$$

Proof. Let $\Delta' = \Delta$ and $h : \Delta' \to \Delta$ the morphism $z \mapsto z^r$. Let X' be the normalization of $X \times_\Delta \Delta'$, $f' : X' \to \Delta'$ the second projection and $X'_0 = (f')^{-1}(0) = \sum d'_j Z'_j$. Local computation gives that $X' \to X$ is étale and $\gcd(d'_j) = 1$. This gives an exact sequence

$$\pi_1(X'_0) \to \pi_1(X_0) \to \mathbb{Z}_r \to 0.$$

Thus it is sufficient to prove that $\pi_1(X_g) \to \pi_1(X'_0)$ is surjective.

From the proof of (2.12) we see the index of the image divides d'_j for every j, thus it is one. ∎

2.13 Proof of (1.12). Let $p : U \to S$ be a flat family of divisors or normal cycles with cycle map $u : U \to X$ on a quasi-projective variety X.

By (2.8.1) there is a dense open $S^0 \subset S$ such that p is a topological fiber bundle over S^0. Thus $\operatorname{im}[\pi_1(U_s) \to \pi_1(X)]$ is independent of $s \in S^0$ (up to conjugation). Repeating this procedure for $S - S^0$ we obtain a stratification of S such that $\operatorname{im}[\pi_1(U_s) \to \pi_1(X)]$ depends only on the stratum containing s. Therefore (1.12) is implied by the following.

2.14 THEOREM. *Let X be a normal quasi-projective variety.*

(2.14.1) [Deligne79] *Let $|H|$ be a pencil with a base point $x \in X$ whose general member H_t is connected. Then $\pi_1(H_t) \to \pi_1(X)$ is surjective.*

(2.14.2) [Campana91] *Let $\{H_t\}$ be a one-parameter family of divisors passing through a point $x \in X$ whose general member H_t is irreducible. Then* $\operatorname{im}[\pi_1(\bar{H}_t) \to \pi_1(X)]$ *has finite index in $\pi_1(X)$.*

Proof. [Campana91] Let $p : W \to S$ be the universal family of divisors $\{H_t\}$ (or of normal cycles $\{\bar{H}_t\}$) with cycle map $w : W \to X$. w is birational in case (2.14.1) and dominant in case (2.14.2). Replacing S by a suitable open subset we may assume that p is a topological fiber bundle. Let $T = w^{-1}(x)$. In case (2.14.1) T is a section of p, which gives a split exact sequence

$$\pi_1(H_t) \to \pi_1(W) \leftrightarrows \pi_1(T) \to 1.$$

By (2.10) $\pi_1(W) \to \pi_1(X)$ is surjective and $\operatorname{im}[\pi_1(T) \to \pi_1(X)] = \{1\}$ since $w(T) = \text{point}$. Thus $\pi_1(H_t) \to \pi_1(X)$ is surjective.

The proof of the second part is similar, but T is only a multisection. As is explained in (3.2), $\pi_1(\bar{H}_t)$ and $\pi_1(T)$ generate a finite index subgroup of $\pi_1(W)$. By (2.10) the image of $\pi_1(W)$ has finite index in $\pi_1(X)$ and $\operatorname{im}[\pi_1(T) \to \pi_1(X)] = \{1\}$. Thus $\operatorname{im}[\pi_1(\bar{H}_t) \to \pi_1(X)]$ has finite index in $\pi_1(W)$. ∎

CHAPTER 3

Shafarevich Maps and Variants

In this chapter I will prove (1.6), (1.8), and some other versions as well.

3.1 Definition. Let G be a group and H_1, H_2 subgroups. We say that H_1 is *essentially a subgroup* of H_2 if $H_1 \cap H_2$ has finite index in H_1. We denote this relationship by $H_1 \lesssim H_2$.

By definition, $H \lesssim \{1\}$ iff H is finite and $G \lesssim H$ iff H has finite index in G.

We need the following easy result:

3.2 PROPOSITION. *Let X be a normal variety. Consider the following diagram:*

$$\begin{array}{ccccccc} T & \xrightarrow{\iota} & V & \xrightarrow{w} & Z & \xrightarrow{n} & X \\ & & \Big\downarrow{\scriptstyle p} & & & & \\ & & V & & & & \end{array}$$

where all the varieties are irreducible and normal, $p \circ i$ and w are dominant. Pick a general point $0 \in T$ and let W_{gen} be the fiber of p containing $i(0)$. For simplicity of notation, all images of 0 are denoted by 0. Then

(3.2.1)

$$\pi_1(Z, 0) \lesssim \langle \mathrm{im}[\pi_1(T, 0) \to \pi_1(Z, 0)], \mathrm{im}[\pi_1(W_{gen}, 0) \to \pi_1(Z, 0)] \rangle.$$

(3.2.2) Let $H \lhd \pi_1(X, 0)$. If

$$\mathrm{im}[\pi_1(T, 0) \to \pi_1(X, 0)] \lesssim H \quad \textit{and}$$
$$\mathrm{im}[\pi_1(W_{gen}, 0) \to \pi_1(X, 0)] \lesssim H,$$

then

$$\mathrm{im}[\pi_1(Z, 0) \to \pi_1(X, 0)] \lesssim H.$$

Proof. By (2.10) $\operatorname{im}[\pi_1(W,0) \to \pi_1(Z,0)]$ has finite index in $\pi_1(Z,0)$; thus it is sufficient to consider the case when $W = Z$. Replacing W by $W \times_V T$ and V by T, we may assume that i is a section $s : V \cong T \to W$. Let V^0 be an open subset and $W^0 = p^{-1}(V^0)$. By suitable choice of V^0, we may assume that $p^0 : W^0 \to V^0$ is a topological fiber bundle (2.8.1). We have a right split exact sequence

$$(3.2.3) \quad \pi_1(W_{gen}, 0) \to \pi_1(W^0, 0) \leftrightarrows \pi_1(V^0, 0) \to 1,$$

which shows that the images of $\pi_1(W_{gen}, 0)$ and of $\pi_1(V^0, 0)$ generate $\pi_1(W^0, 0)$. This proves (3.2.1).

In order to see (3.2.2), let $H_1 = \operatorname{im}[\pi_1(W_{gen}, 0) \to \pi_1(X, 0)]$, $H_2 = \operatorname{im}[\pi_1(V^0, 0) \to \pi_1(X, 0)]$, and $H_3 = \operatorname{im}[\pi_1(W^0, 0) \to \pi_1(X, 0)]$. Then $H_1 \lhd H_3$ and $H_3 = H_1 H_2$ by (3.2.3). $H_1 \lesssim H$ and $H_2 \lesssim H$ since $H_2 \subset \operatorname{im}[\pi_1(T, 0) \to \pi_1(X, 0)]$.

Let $H_i' = H_i \cap H$ and $H_i = \cup_j b_{ij} H_i'$. Since H_1 is normal in H_3,

$$H_3 = H_1(\cup b_{2j} H_2') = \cup b_{2j}(b_{2j}^{-1} H_1 b_{2j}) H_2' = \cup_{j,k} b_{2j} b_{1k} H_1' H_2'.$$

Thus $H_3 \lesssim H$. ∎

For later use we record the following consequence of the proof.

3.2.4 Corollary. *Notation as above. Let T^n (resp. W_{gen}^n) be the normalization of the closure of $w(i(T))$ (resp. $w(W_{gen})$). There are*

(3.2.4.1) finite index subgroups $G_1 < \pi_1(Z, 0)$, $G_2' < \pi_1(T^n, 0)$ and $G_3' < \pi_1(W_{gen}^n, 0)$, and

(3.2.4.2) surjective homomorphisms $G_2' \to G_2$ and $G_3' \to G_3$,
such that we have an exact sequence

$$1 \to G_3 \to G_1 \to G_2 \to 1.$$

Proof. Let $G_1 = \operatorname{im}[\pi_1(W^0, 0) \to \pi_1(Z, 0)]$, $G_2' = \operatorname{im}[\pi_1(V^0, 0) \to \pi_1(T^n, 0)]$ and $G_3' = \operatorname{im}[\pi_1(W_{gen}, 0) \to \pi_1(W_{gen}^n, 0)]$. By (2.10) all these images have finite index in the respective targets. $G_3 = \operatorname{im}[G_3' \to \pi_1(Z, 0)]$ is a normal subgroup in G_1 by (3.2.3). Set

$$G_2 = \operatorname{im}[G_2' \to \pi_1(Z, 0)] / (G_3 \cap \operatorname{im}[G_2' \to \pi_1(Z, 0)]). \quad ∎$$

The above proposition allows us to prove in various situations that through a given very general point of X there is a unique largest cycle satisfying certain properties:

3.3 COROLLARY. *Let X be a normal variety, $x \in VG(X)$ and $H \triangleleft \pi_1(X)$ a normal subgroup. Let $w_1 : W_1 \to X$ be a normal cycle such that $x \in \operatorname{im} w_1$. Let $u : U \to X$ be another normal cycle such that* $\operatorname{im} u$ *intersects* $\operatorname{im} w_1$ *but is not contained in it. Pick base points $0 \in W_1$ and $0 \in U$ such that $w_1(0) = u(0) = 0 \in X$. Assume that* $\operatorname{im}[\pi_1(W_1, 0) \to \pi_1(X, 0)] \lesssim H$ *and* $\operatorname{im}[\pi_1(U, 0) \to \pi_1(X, 0)] \lesssim H$.

Then there is a normal cycle $w_2 : W_2 \to X$ such that $\operatorname{im} w_2 \supset \operatorname{im} w_1$, $\dim W_2 > \dim W_1$, *and* $\operatorname{im}[\pi_1(W_2, 0) \to \pi_1(X, 0)] \lesssim H$.

Proof. By assumption, W_1 is the member of a dominant locally topologically trivial family of cycles $\mathbf{w}_1 : \mathbf{W}_1 \to X$.

Assume first that $\mathbf{w}_1^{-1}(0)$ is positive dimensional at some point of $W_1 \subset \mathbf{W}_1$. Pick a curve $V \subset \mathbf{w}_1^{-1}(0)$ that intersects W_1. Let $W \to V$ be the family induced from $\mathbf{W}_1$ by base change. By construction this has a section. Let W_2 be the normalization of $\operatorname{im}[W \to X]$. By construction $\operatorname{im}[\pi_1(V, 0) \to \pi_1(X, 0)] = \{1\}$. Thus by (3.2), W_2 satisfies the requirements.

Assume next that $\mathbf{w}_1^{-1}(0)$ is zero dimensional along W_1. Then $\mathbf{w}_1$ is quasifinite and dominant near $W_1 \cap \mathbf{w}_1^{-1}(0)$, hence open (see, e.g., [Mumford76, 3.10]). Thus $\mathbf{w}_1^{-1}(\operatorname{im} u)$ contains a curve V that intersects W_1. Construct W_2 as before. In this case $\operatorname{im}[\pi_1(V, 0) \to \pi_1(X, 0)] < \operatorname{im}[\pi_1(U, 0) \to \pi_1(X, 0)]$; thus again (3.2) implies that W_2 satisfies the requirements. ∎

We note the following obvious consequence. (If $H \triangleleft G$ and $g \in G$, then $H_1 \lesssim H$ iff $g^{-1}H_1 g \lesssim H$. Therefore we can ignore the base point in subsequent statements.)

3.4 COROLLARY. *Notation as in (3.3). Pick $x \in VG(X)$. Let $w_0 : W_0 \to X$ be a normal cycle with the following properties:*

(3.4.1) $x \in \operatorname{im} w_0$.

(3.4.2) $\operatorname{im}[\pi_1(W_0) \to \pi_1(X)] \lesssim H$.

(3.4.3) $\dim W_0$ *is the maximal possible satisfying (3.4.1–2).*

Let $u : U \to X$ be any other normal cycle such that $\operatorname{im}[\pi_1(U) \to \pi_1(X)] \lesssim H$. *Then either* $\operatorname{im} u$ *and* $\operatorname{im} w_0$ *are disjoint or* $\operatorname{im} u \subset \operatorname{im} w_0$. *In particular, w_0 is unique.* ∎

3.4.4 COROLLARY. *Notation as in (3.3–4). For any $x \in X$ set*

$$m(x) = \max\{\dim W \mid W \to X \text{ satisfies (3.4.1–2)}\}.$$

If $x \in VG(X)$ and $x' \in X$, then $m(x) \le m(x')$.

Proof. Let $w_0 : W_0 \to X$ be a normal cycle through x such that $\dim W_0 = m(x)$. By definition there is a dominant locally topologically trivial family of normal cycles $S \stackrel{p}{\longleftarrow} W \stackrel{w}{\longrightarrow} X$ extending w_0. Let $C^0 \subset W$ be an irreducible and smooth curve such that x' is contained in the closure of $w(C^0)$. By base change we obtain a locally topologically trivial family of normal cycles

$$C^0 \longleftarrow W \times_S C^0 \longrightarrow X.$$

Let $C \supset C^0$ be a smooth compactification. There is a partial compactification $W_C \supset W \times_S C^0$ such that $p_C \times w_C : W_C \to C \times X$ is finite where $p_C : W_C \to C$ and $w_C : W_C \to X$ are the natural extensions. By construction there is a point $c \in C$ and an irreducible component $Z \subset p_C^{-1}(c)$ such that $x' \in w_C(Z)$. Let $c \in U \subset C$ be a small open neighborhood in the Euclidean topology.

$p_C^{-1}(c)$ is a fiber of $W_C \to C$, hence $\dim Z = \dim W_C - 1 = m(x)$. The natural map $\pi_1(Z) \to \pi_1(X)$ factors through $\pi_1(p_C^{-1}(U)) \to \pi_1(X)$ and by (2.12)

$$\mathrm{im}[\pi_1(p_C^{-1}(U)) \to \pi_1(X)] \lesssim H.$$

Let $Z' \to X$ be the normalization of $\mathrm{im}[Z \to X]$. By (2.10)

$$\mathrm{im}[\pi_1(Z') \to \pi_1(X)] \lesssim \mathrm{im}[\pi_1(Z) \to \pi_1(X)] \lesssim H.$$

Thus $m(x') \geq \dim Z' = m(x)$. ∎

3.5 Definition. Let X be a normal variety and $H \triangleleft \pi_1(X)$ a normal subgroup. A normal variety $\mathrm{Sh}^H(X)$ and a rational map $\mathrm{sh}_X^H : X \dashrightarrow \mathrm{Sh}^H(X)$ are called the *H-Shafarevich variety* and the *H-Shafarevich map* of X if

(3.5.1) sh_X^H has connected fibers, and

(3.5.2) there are countably many closed subvarieties $D_\iota \subset X$ ($D_\iota \neq X$) such that for every closed, irreducible subvariety $Z \subset X$ such that $Z \not\subset \cup D_\iota$.

$$\mathrm{sh}_X^H(Z) = \text{point} \quad \text{iff} \quad \mathrm{im}[\pi_1(\bar{Z}) \to \pi_1(X)] \lesssim H.$$

It is easy to see that $\mathrm{sh}_X^H : X \dashrightarrow \mathrm{Sh}^H(X)$ is unique up to birational equivalence if it exists. If $H = \{1\}$ then we omit the superscript H.

In analogy with (0.3.3) and (1.3), one can define the notion of *H-Shafarevich morphism $sh_X^H : X \to Sh^H(X)$.*

3.5.3 Comment. The map $\pi_1(\bar{Z}) \to \pi_1(X)$ is defined only up to conjugation by elements of the source and target. $\mathrm{im}[\pi_1(\bar{Z}) \to \pi_1(X)]$

is defined up to conjugation by elements of the target. Since H is normal, the condition $\operatorname{im}[\pi_1(\bar{Z}) \to \pi_1(X)] \lesssim H$ is invariant under conjugation.

3.5.4 Remark. If X is defined over a field $F \subset \mathbb{C}$, then clearly $\operatorname{Sh}(X)$ is also defined over F. The same holds for $\operatorname{Sh}^H(X)$ provided H is "defined over F". This notion can be made precise using the Galois action on the algebraic fundamental group [SGA1].

3.6 THEOREM. [Kollár93b, 3.5] *Let X be a normal variety, $H \triangleleft \pi_1(X)$ a normal subgroup. Then,*

(3.6.1) The H-Shafarevich map $\operatorname{sh}_X^H : X \dashrightarrow \operatorname{Sh}^H(X)$ exists.

(3.6.2) For every choice of $\operatorname{Sh}^H(X)$ (within its birational equivalence class) there are open subsets $X^0 \subset X$ and $Z^0 \subset \operatorname{Sh}^H(X)$ with the following properties:

(3.6.2.1) $\operatorname{sh}_X^H : X^0 \to Z^0$ is everywhere defined on X^0.

(3.6.2.2) Every fiber of $\operatorname{sh}_X^H |X^0$ is closed in X.

(3.6.2.3) $\operatorname{sh}_X^H |X^0$ is a topologically locally trivial fibration.

(3.6.3) If X is proper, then $\operatorname{sh}_X^H : X^0 \to Z^0$ is proper and $VG(X) \subset X^0$ for a suitable choice of $\operatorname{Sh}^H(X)$ and X^0.

3.6.4 Comment. This result does not imply that one can choose the D_i in (3.5.2) to be subvarieties of $X - X^0$. For instance, if X is a smooth projective variety that is birational to an Abelian variety A, then we can choose $\operatorname{sh}_X : X \cong X$ as a representative. On the other hand, all the exceptional divisors of $X \to A$ have to be among the D_i.

3.6.5 Example. Let $X' = \mathbb{P}^1 \times \mathbb{C}^*$ with coordinates $(s : t, u)$. Let $C \subset X'$ be the curve $(0:1) \times \mathbb{C}^*$. Blow up the points $(1:1,1), (-1:1,1)$ and contract the birational transform of $\mathbb{P}^1 \times (1)$. We obtain a surface X''; let $C'' \subset X''$ be the birational transform of C. $X'' \to \mathbb{C}^*$ is a proper morphism, the fiber over $u \in \mathbb{C}^*$ is $\mathbb{P}^1$ for $u \neq 1$ and a pair of intersecting lines over $u = 1$. C'' is a section that passes through the singular point of the fiber over 1 (X'' is also singular there). Let $X = X'' - C''$.

It is easy to see that $VG(X) = X$. The family $\mathbb{C} \times (\mathbb{C}^* - \{1\})$ has two extensions to a fiber bundle over $\mathbb{C}^*$: one can use either component of the fiber over 1. The Shafarevich morphism is $X \to \mathbb{C}^*$ but there is no maximal open set over which it is fiber bundle (unless one allows a nonseparated base).

3.6.6 Example. [Blanchard56] Let $\mathcal{O}(a)$ be a line bundle of positive degree over $\mathbb{P}^1$ and M the total space of $\mathcal{O}(a)$. $\mathcal{O}(a)$ is generated by two sections s and t. The four sections

$$(s, t), (is, -it), (t, -s), (it, is) \in \Gamma(\mathcal{O}(a) + \mathcal{O}(a))$$

are independent over $\mathbb{R}$ in every fiber of $\mathcal{O}(a)+\mathcal{O}(a)$. Therefore they generate a lattice bundle $L \subset M+M$, and the quotient $X := M+M/L$ is a compact complex 3-fold such that $\pi_1(X) \cong \mathbb{Z}^4$. There is a natural projection $f : X \to \mathbb{P}^1$, and every fiber is a compact complex torus. (In fact, any fiber is isomorphic to $\mathbb{C}/(\mathbb{Z}+i\mathbb{Z})+\mathbb{C}/(\mathbb{Z}+i\mathbb{Z})$, and is therefore projective.)

Any two points of $M+M$ are connected by a rational curve, thus the same holds for X. There is no Shafarevich morphism for X. There are many ways to see that X is not algebraic and not even bimeromorphic to a Kähler space.

Proof of (3.6). Pick $x \in VG(X)$ and let $w_0 : W_0 \to X$ be as in (3.4). From (2.8) we obtain a locally topologically trivial family of normal cycles

$$S \xleftarrow{\ p\ } W \xrightarrow{\ w\ } X,$$

which contains W_0 such that $\operatorname{im} w_s \neq \operatorname{im} w_{s'}$ for $s \neq s' \in S$. Our aim is to prove that there is an open subset $S^0 \subset S$ such that $w : s^{-1}(S^0) \to X$ is an open immersion.

Choose $x' \in VG(X)$ and let $w_{s(x')} : W_{s(x')} \to X$ be a normal cycle such that $x' \in \operatorname{im} w_{s(x')}$. By (3.4), $m(x) = m(x')$, hence $w_{s(x')} : W_{s(x')} \to X$ is the unique normal cycle whose existence is proved in (3.4). Therefore, if $\operatorname{im} w_{s(x')}$ intersects $\operatorname{im} w_s$ for some $s \in S$, then $s = s(x')$. Thus a very general point of X has only one preimage in W. Therefore w is birational. Let $D_W \subset W$ be the closed subset where w is not an immersion and $D_S = p(D_W)$. $D_S \subset S$ is constructible. If $s \in S$ is very general, then $\operatorname{im} w_s$ does not intersect any other $\operatorname{im} w_{s'}$, hence $s \not\in D_S$. Therefore D_S is not dense in S and we can choose an open subset $S^0 \subset S$ that is disjoint from D_S. Set $W^0 = p^{-1}(S^0)$. By construction $w^0 = w|W^0 : W^0 \to X$ is an open immersion.

Set $X^0 := w^0(W^0)$, $Z^0 := S^0$ and $\operatorname{sh}^H_X := p \circ (w^0)^{-1}$. By definition, $X^0 \subset X$ is open and

$$[\operatorname{sh}^H_X |X^0 : X^0 \to S^0] \cong [p|W^0 : W^0 \to S^0]$$

is topologically locally trivial. The fibers of $\operatorname{sh}^H_X |X^0$ are precisely the normal cycles $W_s \subset X$ for $s \in S^0$, thus they are closed by definiton.

Assume that $u : U \to X$ is a normal cycle such that $\operatorname{im}[\pi_1(U) \to \pi_1(X)] \lesssim H$ and such that $\operatorname{im} u$ intersects a very general fiber W_s of $\operatorname{sh}^H_X |X^0$. If $\operatorname{im} u \not\subset \operatorname{im} w_s$ then we obtain a contradiction by (3.4). Thus $\operatorname{im} u \subset \operatorname{im} w_s$, which shows that (3.5.2) is also satisfied.

If $\phi : \mathrm{Sh}^H(X) \dashrightarrow \mathrm{Sh}^H(X)'$ is another birational model and ϕ is an isomorphism over an open set U, then set $Z^{0\prime} := \phi(Z^0 \cap U)$, $X^{0\prime} := (\mathrm{sh}^H_X)^{-1}(Z^0 \cap U)$.

If X is proper, then there is a maximal choice of X^0 in some sense. In this case any choice of $X^0 \to Z^0$ gives a morphism $Z^0 \to \mathrm{Hilb}(X)$. Let $Z^* \subset \mathrm{Hilb}(X)$ be the union of the images of the various Z^0 and $U^* \to Z^*$ the universal family. Then $u^* : U^* \to X$ is birational and locally an open immersion. Hence u^* is an open immersion. This gives the maximal choice of $X^* := u^*(U^*)$.

In the original construction $x \in X^0$, hence $VG(X) \subset X^*$. ∎

There are some obvious functoriality properties of Shafarevich maps. Proofs are left to the reader (cf. [Kollár93b, 3.6–8]).

3.7 THEOREM. *Let $f : X \to Y$ be a dominant morphism between normal varieties. Let $H \lhd \pi_1(X)$ and $G \lhd \pi_1(Y)$ be normal subgroups. Assume that $f_*H \lesssim G$.*

(3.7.1) There is a rational map $\mathrm{sh}(f) : \mathrm{Sh}^H(X) \dashrightarrow \mathrm{Sh}^G(Y)$, *which makes the following diagram commutative:*

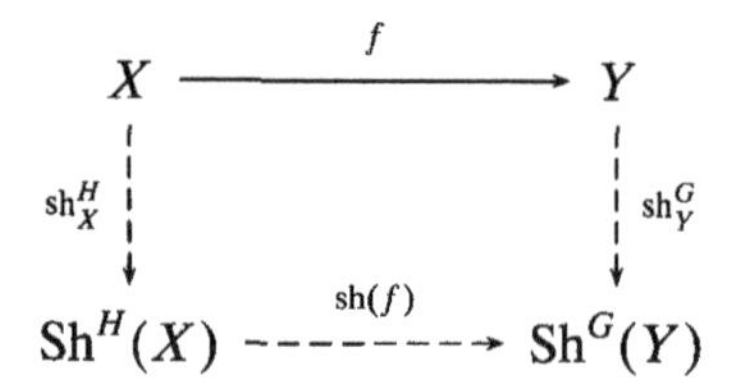

(3.7.2) Assume that $f : X \to Y$ is finite and étale. Then $\mathrm{Sh}(X)$ *is the normalization of* $\mathrm{Sh}(Y)$ *in the function field of X.*

(3.7.3) Assume that X, Y are smooth and proper and f is birational. Then f induces a birational map $\mathrm{sh}(f) : \mathrm{Sh}(X) \dashrightarrow \mathrm{Sh}(Y)$. ∎

We are finally in a position to prove (1.6).

3.8 Proof of (1.6). Let $H = \mathrm{im}[\pi_1(Y) \to \pi_1(X)]$. By (2.11) the normalizer $N(H)$ of H in $\pi_1(X)$ has finite index. Let $X' \to X$ be the corresponding finite étale cover. Since $N(H) \supset H$, the morphism $f : Y \to X$ can be lifted to a morphism $f' : Y \to X'$. (The lifting is not unique; it does not matter which one we choose.) We can identify $\pi_1(X')$ with $N(H)$ and clearly $H = \mathrm{im}[\pi_1(Y) \to \pi_1(X')]$.

H is normal in $\pi_1(X')$ and by (3.6) the H-Shafarevich map

$$\mathrm{sh}^H_{X'} : X' \dashrightarrow \mathrm{Sh}^H(X')$$

exists. Using the notation of (3.6), set $U = (X')^0$, $V = S^0$, and let $g = \mathrm{sh}^H_{X'} |(X')^0$. We obtain $g : U \to V$, which satisfies the conditions (1.6.2.1–2) by (3.6.2). (1.6.2.3) follows from the definition of the H-Shafarevich map. The index of H in $\mathrm{im}[\pi_1(U_v) \to \pi_1(X')]$ is finite by the definition of the H-Shafarevich map. ∎

As was mentioned in (1.7) this construction does not control $\pi_1(U_{gen})$, only its image in $\pi_1(X)$. It is, however, possible to give a variant of the method that gives some control over the fundamental group of the general fiber. It is most convenient to describe it in terms of an abstract condition for a class of groups.

3.9 Definition. Let $\mathcal{F}$ be a class of groups. We say that $\mathcal{F}$ satisfies condition (3.9) if the following statements are valid:

(3.9.1) If $G \in \mathcal{F}$ and $G \to H$ is a surjective homomorphism, then $H \in \mathcal{F}$.

(3.9.2) If $H < G$ is a finite index subgroup, then $H \in \mathcal{F}$ iff $G \in \mathcal{F}$.

(3.9.3) If $1 \to G_1 \to G_2 \to G_3 \to 1$ is exact and $G_1, G_3 \in \mathcal{F}$, then $G_2 \in \mathcal{F}$.

Let $\mathcal{F}$ be a class of groups. Clearly there is a unique smallest class $\overline{\mathcal{F}} \supset \mathcal{F}$, which satisfies condition (3.9).

3.10 Examples. (3.10.1) If H is any finite group and $\mathcal{F} = \{H\}$, then $\overline{\mathcal{F}}$ is the class of all finite groups.

(3.10.2) If H is any finitely generated infinite Abelian group and $\mathcal{F} = \{H\}$, then $\overline{\mathcal{F}}$ is the class of all polycyclic by finite groups (see, e.g., [Segal83, p. 2] for the definition).

(3.10.3) If H has a finite index solvable subgroup and $\mathcal{F} = \{H\}$, then every group in $\overline{\mathcal{F}}$ has a finite index solvable subgroup. (This needs a little argument. See [Segal83, p. 2] for the necessary trick.)

(3.10.4) Let $\mathcal{F}$ be the class of all groups G that have a unique smallest finite index subgroup. Every finite group is in $\mathcal{F}$ but there are finitely generated infinite groups in $\mathcal{F}$ as well.

The following theorem is a generalization of (1.8).

3.11 THEOREM. *Let X be a proper and normal variety and $\mathcal{F}$ a class of groups satisfying (3.9). There is an open set $X^0 \subset X$ and a proper morphism $f^0 : X^0 \to Z^0$ with the following properties:*

(3.11.1) f^0 is a topologically locally trivial fibration.

(3.11.2) For every $z \in Z^0$ the fundamental group of X^0_z is in $\mathcal{F}$.

(3.11.3) If $z \in Z^0$ is very general and $w : W \to X$ is a positive dimensional normal cycle such that $\mathrm{im}\, w$ *intersects X^0_z and $\pi_1(W) \in \mathcal{F}$, then* $\mathrm{im}\, w \subset X^0_z$.

Proof. Choose $H \triangleleft \pi_1(X)$ such that $H \in \mathcal{F}$ and the corresponding Shafarevich map $\mathrm{sh}_X^H : X \dashrightarrow \mathrm{Sh}^H(X)$ has the largest possible fiber dimension. Let $X^* \subset X$ be the open set that was denoted by X^0 in (3.6).

3.11.4 CLAIM. *If $z \in \mathrm{Sh}^H(X)$ is very general and $t : T \to X$ is a positive dimensional normal cycle such that* $\mathrm{im}\, t$ *intersects X_z but* $\mathrm{im}\, t \not\subset X_z$, *then $\pi_1(T)$ is not in $\mathcal{F}$.*

Proof. Assume the contrary. Let $V' = \mathrm{sh}_X^H(X^* \cap t(T))$ and V the normalization of V'. Let W be the normalization of $(\mathrm{sh}_X^H)^{-1}(\mathrm{sh}_X^H(X^* \cap t(T)))$ and Z the normalization of the closure of $(\mathrm{sh}_X^H)^{-1}(\mathrm{sh}_X^H(X^* \cap t(T)))$. $W \to X$ is an open immersion over $W'' = (\mathrm{sh}_X^H)^{-1}(V'')$, where $V'' \subset V$ is the set of normal points.

Let $T^n \subset Z$ be the preimage of $\mathrm{im}\, t$. $T^n \to \mathrm{im}\, t$ is a local isomorphism over $\mathrm{im}\, t \cap W''$, hence $T^n \to \mathrm{im}\, t$ is birational and finite, thus $T^n \cong T$. Apply (3.2.4) to conclude that $\pi_1(W) \in \mathcal{F}$.

By (2.11) and (3.13), the normal subgroup $H_W \triangleleft \pi_1(X)$ generated by $\mathrm{im}[\pi_1(W) \to \pi_1(X)]$ is also in $\mathcal{F}$. $\dim W > \dim X_z$, thus the H_W-Shafarevich map has fiber dimension greater than sh_X^H. This is impossible. ∎

If sh_X^H is birational, then take f^0 to be the identity. If $\mathrm{Sh}^H(X)$ is a point, then H has finite index in $\pi_1(X)$; thus $\pi_1(X) \in \mathcal{F}$ and take $f^0 : X \to \text{point}$.

Let U_z be a very general fiber of sh_X^H. By the above considerations we may assume that $\dim U_z < \dim X$. By induction on the dimension we may assume that there is a map $g^0 : U_z^0 \to Z_U^0$ satisfying the assumptions of (3.11). Pick a point $x \in U_z$ which is very general on X and let $W_x \subset U_z \subset X$ be the fiber of g^0 through x. Choose

$$S \xleftarrow{p} W \xrightarrow{w} X$$

as in the proof of (3.6) such that W_x is a member of W.

3.11.5 CLAIM. *There is a dense open set $S^0 \subset S$ such that $w : p^{-1}(S^0) \to X$ is an open immersion.*

Proof. As in the proof of (3.6), this follows once we establish that if $s \in S$ and $w(W_s)$ intersects W_x, then $W_s = W_x$. By definition $\pi_1(W_s) \in \mathcal{F}$. Thus by (3.11.4) $w(W_s) \subset U_z$. Then $W_s = W_x$ since the assumption (3.11.3) is satisfied by $U_z \dashrightarrow Z_U$. ∎

Set $X^0 = w(p^{-1}(S^0))$ and $f^0 = p \circ w^{-1}|X^0$ to conclude the proof of (3.11). ∎

The proof of (3.11) is very different in nature from the proof of (1.6) given in (3.8). In (3.3) we start with cycles $W \to X$ such that $\operatorname{im}[\pi_1(W) \to \pi_1(X)] \lesssim H$ and gradually build larger dimensional cycles with this property. Finally, the procedure stops when we reach the fibers of the H-Shafarevich map. The following example shows that if one tries to follow the same procedure for (3.11), then one runs into some difficulties: the fundamental group of the intermediate cycles may be too large. This problem can be circumvented but I do not know any simple and clear way to do it.

3.12 Example. Let $X \subset \mathbb{P}^4$ be a smooth cubic hypersurface. It is not hard to see that X is covered by lines and there are 6 lines through a general point. We try to show that $\pi_1(X)$ is finite by using these lines. (In fact, X is simply connected.)

Pick a very general point $x \in X$. Let $l \subset X$ be a line through x. As the natural next step we take $Z \subset X$ to be the union of all lines intersecting l and let $z : \bar{Z} \to X$ be the corresponding normal cycle.

I claim that $\bar{Z}$ is a ruled surface over a curve C' of genus 11. In particular $\pi_1(\bar{Z})$ is infinite.

Let us project X from l to $\mathbb{P}^2$. This exhibits $B_l X$ (the blow up of X along l) as a conic bundle over $\mathbb{P}^2$: a 2-plane containing l intersects X in a cubic curve that contains l. The rest is a conic which is a fiber of the projection.

If we intersect X with a general hyperplane containing l, then we obtain a cubic surface $l \subset H$. On H there are 5 pairs of lines intersecting l. Thus the lines in Z come in pairs and these pairs are parametrized by a smooth plane curve C of degree 5. $\bar{Z}$ is a ruled surface over an étale double cover C' of C. Thus C' has genus 11.

The following group-theoretic result was used in (3.11.4).

3.13 PROPOSITION. *Let G be a group and $H_0 < G$ a subgroup. Assume that $N_G(H_0)$ has finite index in G. Let G_0 be the normal subgroup of G generated by H_0. If $\mathcal{F}$ satisfies (3.9) and $H_0 \in \mathcal{F}$, then $G_0 \in \mathcal{F}$.*

Proof. We may as well assume that $G = G_0$. By (3.15) there is a finite index subgroup $G^0 < G$ such that G^0 is generated by normal subgroups H_i and each H_i is isomorphic to a finite index subgroup of H_0. In particular, $H_i \in \mathcal{F}$. G^0 has a composition series whose successive quotients are

$$H_1 \cdots H_m / H_1 \cdots H_{m-1} \cong H_m / H_m \cap H_1 \cdots H_{m-1}. \quad \blacksquare$$

3.14 LEMMA. *Let G be a group and $S = \{s_1, \ldots, s_k\} \subset G$ a set of generators, invariant under conjugation. Then every element $g \in G$ can be written in the form*

$$g = s_1^{n_1} s_2^{n_2} \cdots s_k^{n_k}, \quad (n_i \in \mathbb{Z}).$$

Thus G is finite if all the s_i are torsion elements. ∎

3.15 LEMMA. *Let G be a group and $H_i < G$ subgroups. Assume that $N_G(H_i)$ has finite index in G for every i. Assume that the normal subgroup G_1 generated by the H_i has finite index in G. Then*

(3.15.1) there is a finite index normal subgroup $G^0 \triangleleft G$ such that G^0 is generated by finitely many normal subgroups $H_{ij}^0 \triangleleft G^0$, where H_{ij}^0 is conjugate to a finite index subgroup of H_i;

(3.15.2) if every H_i is finite, then G is finite.

Proof. By assumption every H_i has only finitely many conjugates H_{ij}.

Assume that every H_i is finite. Set $S = \cup H_{ij}$. S satisfies the assumptions of (3.14), thus G_1 and G are finite.

Let $G^1 = \cap_{ij} N_G(H_{ij})$. G^1 is a finite index normal subgroup of G. G^1 normalizes every H_{ij}; thus $H_{ij}^0 = H_{ij} \cap G^1$ is normal in G^1 and has finite index in H_{ij}. Let G^0 be the subgroup of G generated by the H_{ij}^0. G^0 is normal in G and the H_{ij}^0 are normal in G^0. We need to show that G/G^0 is finite.

The subgroups $H_{ij}/(H_{ij} \cap G^0) < G_1/G^0$ are finite and they generate G_1/G^0. Thus by the second part G_1/G^0 is finite, hence so is G/G^0. ∎

Finally we describe the relationship between the original Shafarevich conjecture and Shafarevich morphisms and maps.

3.16 Notation. Let X be a normal complex space and $H \triangleleft \pi_1(X)$ a normal subgroup. Let $\rho^H : X^H \to X$ denote the étale Galois cover with Galois group $\pi_1(X)/H$.

3.17 PROPOSITION. *Let $f : X \to Y$ be a proper morphism of normal complex spaces and $H \triangleleft \pi_1(X)$ a normal subgroup. Assume that if $Z \subset X$ is a connected subspace such that $f(Z) =$ point, then* $\operatorname{im}[\pi_1(Z) \to \pi_1(X)] \lesssim H$. *Then there is a commutative diagram*

$$\begin{array}{ccc} X^H & \xrightarrow{f^H} & Y^H \\ \downarrow{\scriptstyle \rho^H} & & \downarrow{\scriptstyle \tau^H} \\ X & \xrightarrow{f} & Y \end{array}$$

where f^H is proper with connected fibers, Y^H is normal, and τ^H has discrete fibers.

Proof. It is clear that Y^H is unique, if it exists. Therefore it is sufficient to construct it locally over small open sets $U \subset Y$.

Pick any point $y \in Y$ and choose $y \in U \subset Y$ such that $f^{-1}(y)$ is a deformation retract of $f^{-1}(U)$. Let $H' \triangleleft \pi_1(f^{-1}(U))$ be the preimage of H. H' has finite index in $\pi_1(f^{-1}(U))$; let $V \to f^{-1}(U)$ be the corresponding cover. The composite $V \to f^{-1}(U) \to U$ is proper, thus it has a Stein factorization

$$\begin{array}{ccc} V & \longrightarrow & U^H \\ \downarrow & & \downarrow \\ f^{-1}(U) & \xrightarrow{f} & U. \end{array} \tag{3.17.1}$$

$(f \circ \rho^H)^{-1}(U)$ is a disjoint union of copies of V, thus (3.17.1) gives τ^H above U. ∎

3.18 THEOREM. *Let X be a normal and proper variety, $H \triangleleft \pi_1(X)$ a normal subgroup. The following are equivalent:*

(3.18.1) The H-Shafarevich morphism $sh_X^H : X \to Sh^H(X)$ exists.

(3.18.2) The corresponding cover $X^H \to X$ admits a proper morphism $sh^H : X^H \to Sh^H(X^H)$ onto a complex space $Sh^H(X^H)$ which does not have any positive dimensional compact complex subspaces.

Proof. Assume that $sh_X^H : X \to Sh^H(X)$ exists. Set $Y = Sh^H(X)$, $f = sh_X^H$, and apply (3.17) to construct

$$X^H \xrightarrow{f^H} Y^H \xrightarrow{\tau^H} Sh^H(X).$$

Let $Z \subset Y^H$ be compact and connected. Then $(f^H)^{-1}(Z) \subset X^H$ is also compact and connected, hence

$$\mathrm{im}[\pi_1(Z') \to \pi_1(X)] < H,$$

where $Z' \subset (f^H)^{-1}(Z)$ is any irreducible component. By (2.10, 3.5) the image of Z' in $Sh^H(X)$ is a point, hence Z is a point.

Conversely, assume that $sh^H : X^H \to Sh^H(X^H)$ exists. By Stein factorization we may assume that it has connected fibers and $Sh^H(X^H)$ is normal. Then sh^H is unique; hence the $\pi_1(X)/H$-action on X^H descends to an action on $Sh^H(X^H)$. The corresponding quotient is $Sh^H(X)$. ∎

A similar result holds for Shafarevich maps. The key point is the following result.

3.19 THEOREM. *Let $f : X \to Y$ be a morphism between proper and smooth varieties with connected general fiber X_g. Let $H \triangleleft \pi_1(X)$ be a normal subgroup and assume that* $\operatorname{im}[\pi_1(X_g) \to \pi_1(X)] \lesssim H$. *Then there are birational models $f' : X' \to Y'$ such that* $\operatorname{im}[\pi_1(Z) \to \pi_1(X')] \lesssim H$ *for every connected subvariety $Z \subset X'$ such that $f'(Z) = point$.*

Proof. There is an open subset $Y^0 \subset Y$ such that f is flat above Y^0. [illegible] morphism $p : Y^0 \to \operatorname{Hilb}(X)$. Choose a smooth compactifi- [illegible] that p extends to a morphism $Y' \to \operatorname{Hilb}(X)$. Let [illegible] rmalization of $Y' \times_{\operatorname{Hilb}(X)} \operatorname{Univ}(X)$ and $X' \to X''$ a [illegible] /e obtain the following diagram:

$$\begin{array}{ccc} X'' & \xleftarrow{q} & X' \\ {\scriptstyle f''}\downarrow & & \downarrow{\scriptstyle f'} \\ Y'' & = & Y \end{array}$$

Pick $Z \subset X'$ such that $f'(Z) = $ point. We need to show that $\operatorname{im}[\pi_1(Z) \to \pi_1(X')] \lesssim H$. Since $\pi_1(X') = \pi_1(X)$, this is equivalent to

$$\operatorname{im}[\pi_1(Z) \to \pi_1(q(Z)) \to \pi_1(X'') \to \pi_1(X)] \lesssim H.$$

The general fiber of f'' is X_g, hence by (2.12)

$$\operatorname{im}[\pi_1(q(Z)) \to \pi_1(X'')] \lesssim \operatorname{im}[\pi_1(X_g) \to \pi_1(X'')].$$

Pushing this forward to X we obtain that

$$\operatorname{im}[\pi_1(Z) \to \pi_1(X)] \lesssim \operatorname{im}[\pi_1(X_g) \to \pi_1(X)] \lesssim H. \quad \blacksquare$$

An argument as in (3.18) gives the following.

3.20 COROLLARY. *Let X' be a smooth and proper variety and $H \triangleleft \pi_1(X')$ a normal subgroup. There is a smooth and proper variety X birational to X' such that the corresponding cover $X^H \to X$ admits a proper morphism* $\operatorname{sh}^H : X^H \to \operatorname{Sh}^H(X^H)$ *onto a complex space* $\operatorname{Sh}^H(X^H)$ *which does not have any positive dimensional compact complex subspaces through a very general point.* $\blacksquare$

CHAPTER 4

The Fundamental Group and the Classification of Algebraic Varieties

In this chapter I will outline the relationship between the fundamental group of an algebraic variety and other properties and invariants.

If we take the product of a variety X_1 with a simply connected variety X_2, then the fundamental group of $X_1 \times X_2$ does not carry any information about X_2. Before we can say anything substantial, we must decide when to expect the fundamental group to govern global properties of the whole variety.

The general outlines of this problem are clear, but it is difficult to pin down the best version at the moment. We use several variants. There are three main cases and each has four subcases. Six of these possibilities are listed here, another six will be added in (4.5). The most natural conditions are (4.1.1–3). These, however, have the problem that they are not birational invariants and are too restrictive unless we consider normal varieties as well. The versions denoted by (4.1.1–3.gen) pose restrictions only on "very general" subvarieties. These are much easier to handle.

4.1 Various notions of "large fundamental group."

(4.1.1) The universal covering of X is Stein.

(4.1.1.gen) The universal covering of X admits a proper bimeromorphic morphism onto a Stein space.

(4.1.2) $\operatorname{im}[\pi_1(W) \to \pi_1(X)]$ is infinite for every positive dimensional normal cycle $w : W \to X$.

(4.1.2.gen) $\operatorname{im}[\pi_1(W) \to \pi_1(X)]$ is infinite for every positive dimensional normal cycle $w : W \to X$ such that $\operatorname{im} w$ intersects $VG(X)$.

(4.1.3) $\pi_1(W)$ is infinite for every positive dimensional normal cycle $w : W \to X$.

(4.1.3.gen) $\pi_1(W)$ is infinite for every positive dimensional normal cycle $w : W \to X$ such that $\operatorname{im} w$ intersects $VG(X)$.

4.1.4 Comment. We have the following implications between these properties:

$$\begin{array}{ccccc} (4.1.1) & \Rightarrow & (4.1.2) & \Rightarrow & (4.1.3) \\ \Downarrow & & \Downarrow & & \Downarrow \\ (4.1.1.\text{gen}) & \Rightarrow & (4.1.2.\text{gen}) & \Rightarrow & (4.1.3.\text{gen}) \end{array}$$

4.2 Algebraic fundamental group. In algebraic geometry it is very difficult to see the fundamental group itself. One can, however, see finite degree coverings of a variety. These correspond to finite index subgroups of the fundamental group.

Given a group G, let $K(G) \subset G$ denote the intersection of all finite index subgroups of G. We say that G is *residually finite* if $K(G) = \{1\}$. If X is any topological space, then $K(\pi_1(X))$ cannot be detected by studying finite covers of X. The first example of a smooth projective algebraic variety such that $K(\pi_1(X)) \neq \{1\}$ was given in [Toledo93]. Further examples were constructed by [Catanese-Kollár92] and by Nori (unpublished).

The algebraic fundamental group of X is essentially the quotient $\pi_1(X)/K(\pi_1(X))$. For the purpose of these notes, this viewpoint does not lead to any problems. In general it is better to take the completion of $\pi_1(X)/K(\pi_1(X))$ in the topology given by finite index subgroups. This way we obtain a compact totally disconnected topological group, denoted by $\hat{\pi}_1(X)$. ([SGA1] defines $\hat{\pi}_1(X)$ for any scheme X.)

4.3 Algebraic Shafarevich map. Let X be a normal quasi-projective variety and $H = K(\pi_1(X))$. H is a normal subgroup. The corresponding H-Shafarevich map is called the *algebraic Shafarevich map*. It is denoted by

$$\widehat{\mathrm{sh}}_X : X \dashrightarrow \widehat{\mathrm{Sh}}(X).$$

4.4 Covering spaces. Given a topological space X, let $\tilde{X}$ be the universal covering space of X. Let $\hat{X} \to X$ be the covering of X corresponding to $K(\pi_1(X)) < \pi_1(X)$. This is called the *universal algebraic covering* of X. (Usually $\hat{X}$ is not an algebraic variety.)

4.5 Various notions of "large algebraic fundamental group." Each of the properties (4.1) have their algebraic counterparts. These are denoted by (4.1.1.alg) etc.

(4.5.1) In case (4.1.1) we can consider the universal algebraic covering of X. In (4.1.2–3) we can use $\hat{\pi}_1$ instead of π_1.

(4.5.2) Each of the six properties in (4.1) is implied by the algebraic version.

This is clear for the last four. In order to see the first two, observe that $\tilde{X}$ is the universal covering of $\hat{X}$. If $\hat{X}$ is Stein, then $\tilde{X}$ is also Stein [Stein56].

(4.5.3) The implications given in (4.1.4) hold for the algebraic versions as well.

We name the two most frequently used variants:

4.6 Definition. Let X be a normal variety. We say that X has *generically large algebraic fundamental group* (resp. *generically large fundamental group*) if X satisfies the condition (4.1.2.alg.gen) (resp. (4.1.2.gen)).

Let X be any variety and $Z \subset X$ a very general fiber of the Shafarevich map. Then $\mathrm{im}[\pi_1(Z) \to \pi_1(X)]$ is finite and Z is maximal with this property. Thus one should expect that $\mathrm{Sh}(X)$ has generically large fundamental group. (Choose a smooth projective model of $\mathrm{Sh}(X)$ in order to talk about $\pi_1(\mathrm{Sh}(X))$.) This is unfortunately false in general.

4.7 Example. Let C be a hyperelliptic curve with hyperelliptic involution τ and S a $K3$ surface with a fixed-point free involution σ. Let $X = C \times S/(\tau, \sigma)$. The Shafarevich map is the natural morphism $X \to C/\tau \cong \mathbb{P}^1$ and the general fiber is S. Thus $\mathrm{im}[\pi_1(S) \to \pi_1(X)] = 1$ and $\pi_1(\mathrm{Sh}(X)) = 1$, but $\pi_1(X)$ is infinite.

In many cases the above pathology can be eliminated by taking a finite étale cover.

4.8 THEOREM. [Kollár93b, 4.5] *Let X be a smooth and proper variety.*

(4.8.1) There is a finite étale cover $X' \to X$ such that $\widehat{\mathrm{sh}}_* : \hat{\pi}_1(X') \to \hat{\pi}_1(\widehat{\mathrm{Sh}}(X'))$ is an isomorphism and $\widehat{\mathrm{Sh}}(X')$ has generically large algebraic fundamental group.

(4.8.2) Assume in addition that $\pi_1(X)$ is residually finite. Then there is a finite étale cover $X' \to X$ such that $\mathrm{sh}_* : \pi_1(X') \to \pi_1(\mathrm{Sh}(X'))$ is an isomorphism, and $\mathrm{Sh}(X')$ has generically large fundamental group. ∎

For the rest of the chapter we consider how the fundamental group fits into the usual rough classification of algebraic varieties. We start with a quick review of the surface case. See [BPV84] for a more detailed account.

4.9 Rough classification of algebraic surfaces. There are five classes:

(4.9.1) Rational surfaces. These are all birational to $\mathbb{P}^2$. Any two points of a rational surface can be connected by a chain of rational curves.

(4.9.2) Irrational ruled surfaces. S admits a map $S \to C$ such that $g(C) > 0$ and every fiber is a union of rational curves.

For both of these cases, $h^0(S, \mathcal{O}_S(mK_S)) = 0$ for $m \geq 1$.

(4.9.3) Abelian surfaces, K3 surfaces and their quotients by finite groups acting freely. For all of these, $h^0(S, \mathcal{O}_S(mK_S)) \leq 1$ for $m \geq 1$ and equality holds for some value of m.

(4.9.4) Elliptic surfaces. For large $m \gg 1$ the sections of $\mathcal{O}_S(mK_S)$ give a morphism $\phi_m : S \to C$ such that general fibers are elliptic curves.

(4.9.5) Surfaces of general type. For large $m \gg 1$ the sections of $\mathcal{O}_S(mK_S)$ give a birational morphism $\phi_m : S \to S'$.

Since I know the rest of the story, it is clear which properties should be emphasized in general.

4.10 Definition. Let X be a proper variety.

(4.10.1) X is called *rationally chain connected* if any two points of X can be connected by a chain of rational curves.

(4.10.2) X is called *rationally connected* if any two general points of X can be connected by an irreducible rational curve

(4.10.3) X is called *uniruled* if there is a rational curve through any point of X.

4.11 Definition. Let X be a smooth and proper variety. We define the *Kodaira dimension* of X (denoted by $\kappa(X)$) as follows:

(4.11.1) $\kappa(X) = -\infty$ iff $h^0(X, \mathcal{O}_X(mK_X)) = 0$ for $m \geq 1$.

(4.11.2) $\kappa(X) = k$ for $k \in \{0, \dots, \dim X\}$ if all the nonzero values of

$$h^0(X, \mathcal{O}_X(mK_X))/m^k$$

are bounded from below and above by positive constants.

It is not hard to see that every variety has a Kodaira dimension [Iitaka71, Ueno75].

4.12 Rough classification of smooth algebraic varieties. There are five classes:

(4.12.1) Rationally connected varieties. For smooth varieties this is equivalent to being rationally chain connected [KoMiMo92]. By [Campana91] if X is rationally connected then $h^i(X, \mathcal{O}_X) = 0$ for $i > 0$, thus $\chi(X, \mathcal{O}_X) = 1$. Also, this notion is deformation invariant [KoMiMo92].

(4.12.2) Uniruled but not rationally connected. By [Campana81; KoMiMo92] there is a sequence of maps

$$X = X_1 \dashrightarrow X_2 \dashrightarrow \dots \dashrightarrow X_k = Z$$

such that Z is not uniruled and every fiber of $X_i \dashrightarrow X_{i+1}$ is rationally connected. Conjecturally we need only one map $X \dashrightarrow Z$.

For both of these cases, $h^0(X, \mathcal{O}_X(mK_X)) = 0$ for $m \geq 1$. Conjecturally the converse also holds: X is uniruled iff $\kappa(X) = -\infty$. (True if $\dim X \leq 3$ by [Miyaoka88].)

(4.12.3) Varieties of Kodaira dimension zero. This is a big class and any classification is unlikely. Several thousand families known in dimension 3. Maybe infinitely many families?

(4.12.4) $0 < \kappa(X) < \dim X$. For large $m \gg 1$ the sections of $\mathcal{O}_X(mK_X)$ give a map $\phi_m : X \dashrightarrow Z$ such that general fibers have Kodaira dimension zero.

(4.12.5) Varieties of general type. For large $m \gg 1$ the sections of $\mathcal{O}_X(mK_X)$ give a birational map $\phi_m : X \dashrightarrow X'_m$. This is the largest class and again very little is known about it.

The fundamental group of a variety of nongeneral type is rather special. In the rest of the chapter I formulate some theorems and conjectures which make this assertion more precise.

4.13 THEOREM. *Let X be a normal and proper variety. Assume that X is rationally chain connected. Then*

(4.13.1) $\pi_1(X)$ is finite;

(4.13.2) [Campana91; KoMiMo92, 2.5] *if X is smooth then $\pi_1(X) = \{1\}$.*

Proof. Let $f : X^0 \to Z$ be as in (1.8). Let $z_1, z_2 \in Z$ be two general points and pick $x_i \in f^{-1}(z_i)$ to be general. By assumption there is a chain of rational curves $C_1, \ldots, C_k$ connecting x_1 and x_2. Applying (1.8.3) to C_1 we obtain that $C_1 \subset f^{-1}(z_1)$. Thus C_2 intersects $f^{-1}(z_1)$ and so $C_2 \subset f^{-1}(z_1)$, and so on. Finally we obtain that $x_2 \in f^{-1}(z_1)$ hence $z_1 = z_2$. Hence by (1.8) $\pi_1(X)$ is finite.

If $\pi_1(X)$ is finite then the universal cover $\tilde{X}$ is also an algebraic variety. It is easy to see that $\tilde{X}$ is again rationally chain connected. As was noted in (4.12.1) this implies that $\chi(X, \mathcal{O}) = \chi(\tilde{X}, \mathcal{O}) = 1$ if X is smooth. On the other hand, $\chi(\tilde{X}, \mathcal{O}) = \deg(\tilde{X}/X)\chi(X, \mathcal{O})$, which shows that $\tilde{X} = X$. ∎

4.14 Example. Let E be an elliptic curve and S the surface $\mathbb{P}^1 \times E$. Let τ be an order two translation on E and consider the automorphism of S given by $\sigma(x, e) = (x^{-1}, \tau e)$. σ is an involution without fixed points.

Fix a point $0 \in E$ and let S' be obtained from S by blowing up the four points

$$(0, 0),\ (0, \tau 0),\ (\infty, 0),\ (\infty, \tau 0).$$

Let C_0 and C_∞ be the birational transforms of the curves $\{0\} \times E$ and $\{\infty\} \times E$ in S'. C_0 and C_∞ can be contracted; this way we obtain a surface S'' with two sigular points and a fixed point free involution σ''. Furthermore, S'' is rationally chain connected. Thus $X = S''/\sigma''$ is a rationally chain connected surface and $\pi_1(X) = \mathbb{Z}_2$.

4.15 THEOREM. [Kollár93b, 5.3]. *Let $f : X \dashrightarrow Z$ be as in (4.12.2). Assume that X and Z are smooth. Then $f_* : \pi_1(X) \to \pi_1(Z)$ is an isomorphism.* ∎

4.16 CONJECTURE. *Let X be a smooth and proper variety with $\kappa(X) = 0$.*

(4.16.1) Then X has a finite étale cover X' such that X' is birational to the product of an Abelian variety and of a simply connected variety with $\kappa = 0$.

(4.16.2) In particular, $\pi_1(X)$ has a finite index Abelian subgroup.

4.17 Remarks.. (4.17.1) If $c_1(X) = 0$ then this is true (see, e.g., [Beauville83]).

(4.17.2) Assume that the Minimal Model Program works (see, e.g., [Kollár87a; CKM88] for introductions). Then (4.16.2) implies (4.16.1) by [Kollár93b, 5.6].

(4.17.3) The following approach works in dimension 3.

Let X be a threefold with $\kappa(X) = 0$. We would like to prove that $\pi_1(X)$ has a finite index Abelian subgroup.

By the Minimal Model Program, X is birational to a variety X' such that $K_{X'}$ is numerically trivial, in fact some multiple of $K_{X'}$ is torsion in $\mathrm{Pic}(X')$. X' has terminal singularities, hence by [Kollár93b, 6.8.1] $\pi_1(X) \cong \pi_1(X')$.

There is a finite (in general not étale) cover $X'' \to X'$ such that $K_{X''} = 0$. Thus $\pi_1(X'')$ maps to a finite index subgroup of $\pi_1(X')$.

$K_{X''} = 0$ implies that X has only cDV singularities; let X''' be a small $\mathbb{Q}$-factorialization. Assume that X''' is smoothable. Let $f : Y \to \Delta$ be a smoothing, $Y_0 \cong X'''$. The general fiber Y_g has trivial K and $\pi_1(Y_g) \to \pi_1(X''')$ is surjective (cf. [Kollár93b, 5.2.2]). Y_g is smooth and $c_1(Y_g) = 0$, thus $\pi_1(Y_g)$ has a finite index Abelian subgoup.

In dimension 3 [Namikawa93/94] showed the smoothability of X'''. Hopefully his approach can be generalized.

(4.17.4) [Mok92] shows that $\pi_1(X)$ does not have any discrete and Zariski dense representation in any semisimple algebraic group.

For larger Kodaira dimension the situation should again be simple.

4.18 CONJECTURE. *Let X be a smooth proper variety with generically large fundamental group such that $0 < \kappa(X) < \dim X$. Then X has a finite étale cover $p : X' \to X$ such that X' is birational to a smooth family* *of Abelian varieties over a projective variety of general type Z which has generically large fundamental group.*

(4.16) implies (4.18):

4.19 THEOREM. [Kollár93b, 6.3] *Let $f : X \to Y$ be a dominant morphism with connected fibers between smooth and proper varieties. Let $y \in Y$ be a very general point. Assume that X has generically large fundamental group, and X_y has a finite étale cover which is birational to an Abelian variety.*

Then X has a finite étale cover $X' \to X$ such that $f' : X' \to Y'$ is birational to a smooth family of Abelian varieties over a proper base, where $X' \to Y' \to Y$ is the Stein factorization of $X' \to X \to Y$. ∎

Part II

Automorphic Forms: Classical Theory

CHAPTER 5

The Method of Poincaré

Let C be a Riemann surface and $p : M \to C$ its universal covering space. M is a simply connected Riemann surface, thus it is either $\mathbb{CP}^1$, $\mathbb{C}$ or the unit disc Δ. In all three cases we understand the function theory of M quite well. The aim of the theory of automorphic forms is to connect the function theory of M and the function theory of C. There are several ways to do this and we explore some of the possibilities.

5.1 Naive approach. Let C be a curve of genus at least 2 with universal cover Δ. Let Γ be the fundamental group of C. Γ acts on Δ by deck transformations. It is clear that there is a one-to-one correspondence between holomorphic (resp. meromorphic) functions on C and Γ-invariant holomorphic (resp. meromorphic) functions on Δ.

C is compact, thus it does not carry any nonconstant holomorphic functions. Therefore we need to look at meromorphic functions on C which correspond to Γ-invariant meromorphic functions on Δ. Meromorphic functions on Δ can be written as quotients of two holomorphic functions g_1, g_2. Unfortunately, the g_i cannot be Γ-invariant, thus again we lose the connection with the group action. The key observation clarifying the problem was made in [Poincaré1883]. (See, e.g., [Gunning76] for a modern treatment and for further results.)

5.2 Poincaré automorphic forms. Instead of trying to find Γ-invariant holomorphic functions (which do not exist), we search for holomorphic functions whose *zero set* is Γ-invarant. If $g(z)$ is such a function then $g(\gamma z)$ and $g(z)$ have the same zero set, thus $J(\gamma, z) = g(\gamma z)/g(z)$ is nowhere zero and holomorphic on Δ. The collection of all these functions $\{J(\gamma, z)\}$ is called the *factor of automorphy*. By computing $g(\gamma_1\gamma_2 z)$ two ways, one sees that the factor of automorphy satisfies the so-called *cocycle condition*:

(5.2.1) $\quad J(\gamma_1\gamma_2, z) = J(\gamma_1, \gamma_2 z)J(\gamma_2, z).$

Let $\{J(\gamma, z)\}$ be a collection of nowhere zero holomorphic functions on Δ satisfying the cocycle condition (5.2.1). A holomorphic function g on

Δ is called an *automorphic form (of type J)* if it satisfies the functional equation

$$(5.2.2) \quad g(\gamma z) = J(\gamma, z)g(z) \quad \forall\, z \in \Delta, \gamma \in \Gamma.$$

It is easy to construct Γ-invariant meromorphic functions using automorphic forms. Assume that g_1 and g_2 are automorphic forms of type J. Then

$$\frac{g_1(\gamma z)}{g_2(\gamma z)} = \frac{J(\gamma, z)g_1(z)}{J(\gamma, z)g_2(z)} = \frac{g_1(z)}{g_2(z)},$$

thus g_1/g_2 is a Γ-invariant meromorphic function.

Poincaré observed that if f is any function on Δ then the series

$$(5.2.3) \quad P(f)(z) = \sum_{\gamma\in\Gamma} J(\gamma, z)^{-1} f(\gamma z)$$

satisfies (5.2.2), provided it converges absolutely. The series (5.2.3) is called the *Poincaré series* of f (of type J).

5.2.4 Examples of factors of automorphy.

(5.2.4.1) The simplest construction is the following. Let f be any invertible holomorphic function and set $T_f(\gamma, z) = f(\gamma z)/f(z)$. T_f is a factor of automorphy. If g is an automorphic form of type T_f then g/f is Γ-invariant and holomorphic, hence constant. Such a factor of automorphy is called *trivial.*

(5.2.4.2) Let $\rho : \Gamma \to \mathbb{C}^*$ be a representation and set $T_\rho(\gamma, z) = \rho(\gamma)$. T_ρ is a factor of automorphy. One can see that the space of automorphic forms of type T_ρ is at most one dimensional. (At least it is clear that a Poincaré series of type T_ρ is always divergent.) Thus again we fail to get any Γ-invariant meromorphic functions. Such a factor of automorphy is called *flat.*

(5.2.4.3) A more fruitful observation is that condition (5.2.1) looks like the chain rule, and indeed

$$J(\gamma, z)^{-1} = \frac{\partial \gamma z}{\partial z}$$

is a factor of automorphy. (The inverse is there to achieve greater conformity with the traditional notation.)

(5.2.4.4) These essentially exhaust all possible factors of automorphy on Δ. It is a nontrivial result (see, e.g., [Gunning76]) that a suitable power of every factor of automorphy can be written in the form $J^k T_f T_\rho$

for suitable $k \in \mathbb{Z}$, f and ρ. There is a unique representation if we assume that ρ is unitary.

5.2.5 Convergence of Poincaré series. Poincaré proved that if f is a bounded holomorphic function then the Poincaré series of f of type J^k is convergent for $k \geq 2$ where J is as in (5.2.4.3). Thus we obtain plenty of automorphic forms, and in fact any meromorphic function on C can be written as the quotient of two such Poincaré series.

5.3 Reinterpretation. We can recast the work of Poincaré in more modern language the following way. A function on Δ is a section of the trivial line bundle $\Delta \times \mathbb{C}$. The Γ action on Δ can be lifted to an action on $\Delta \times \mathbb{C}$ by acting trivially on $\mathbb{C}$. There can be other liftings as well. Any lifting has the form

$$(5.3.1) \quad \gamma(z, t) = (\gamma z, j(\gamma, z)t).$$

It is easy to check that this is a group action iff $j(\gamma, z)$ satisfies the condition

$$(5.3.2) \quad j(\gamma_1\gamma_2, z) = j(\gamma_1, \gamma_2 z)j(\gamma_2, z).$$

It is now clear that we should take $j(\gamma, z) = J(\gamma, z)$ to obtain a correspondence between factors of automorphy as in (5.2.1) and liftings of the Γ action.

What happens with the Poincaré series in this way? f is still a section of the trivial line bundle, but we changed the identification between the stalks $\mathbb{C}_z$ and $\mathbb{C}_{\gamma z}$. Strictly speaking, $f(\gamma z)$ and $f(z)$ take up values in different vector spaces, and our favorite identification is now given by γ. Compared with the old trivialization, this is just multiplication by the factor $J(\gamma, z)$. Thus the Poincaré series becomes

$$P(f)(z) = \sum_{\gamma \in \Gamma} f(\gamma z).$$

5.4 General form. More generally, we can try to consider any line bundle L_M on M such that the Γ-action on M lifts to a Γ-action on L_M. If $f \in H^0(M, L_M)$ is a global section, then we can consider

$$(5.4.1) \quad P(f)(x) = \sum_{\gamma \in \Gamma} f(\gamma x).$$

$P(f)$ is a Γ-invariant section of L_M, called the *Poincaré series* associated to f. Assuming convergence, P can be viewed as a linear operator

$$P : H^0(M, L_M) \to H^0(M, L_M)^\Gamma,$$

where, as usual, a superscript Γ indicates the Γ-invariant elements of a set.

5.4.2 Example. Let M be a complex manifold. If $p : M \to M$ is a biholomorphism, then we obtain a natural isomorphism $p_* K_M \cong K_M$. This defines an action of $\operatorname{Aut}(M)$ on K_M.

If $M \subset \mathbb{C}^n$ is open, then $dz_1 \wedge \cdots \wedge dz_n$ is a nowhere zero section of K_M and the corresponding factor of automorphy is

$$J(\gamma, z)^{-1} = \frac{\gamma^* dz_1 \wedge \cdots \wedge \gamma^* dz_n}{dz_1 \wedge \cdots \wedge dz_n}.$$

This is the determinant of the Jacobian matrix of the transformation $\gamma : M \to M$, a higher-dimensional version of (5.2.4.3).

The arguments of Poincaré can be generalized to a more general setting. We start with very few assumptions.

5.5 Definition. Let M be a topological space and Γ a group of homeomorphisms of M. We say that Γ acts *discontinuously* on M if the following two conditions are satisfied:

(5.5.1) For every $x \in M$ the stabilizer $\operatorname{Stab}(x) < \Gamma$ is finite.

(5.5.2) For every $x \in M$ there is an open neighborhood $x \in U \subset M$ such that $U \cap \gamma U = \emptyset$ iff $x \neq \gamma x$.

5.6 Definition. (5.6.1) The action is called *fixed point free* (or *free*) if $x = \gamma x$ implies $\gamma = 1$.

(5.6.2) A subset $D \subset M$ is called a *fundamental domain* if $M = \cup \overline{\gamma D}$ and $\overline{\gamma D}$ is disjoint from the interior of D for $\gamma \neq 1$. (I usually also assume that D is the closure of its interior and the boundary of D is "small," for instance it has measure zero.)

(5.6.3) We say that Γ has a *compact fundamental domain* if there is a compact set $K \subset M$ such that $\cup_{\gamma \in \Gamma} \gamma K = M$. (For discontinuous actions this is equivalent to having a fundamental domain that is compact. The action of $\mathbb{Q}$ on $\mathbb{R}$ by translations shows that these are not always equivalent.)

The following result is easy. We mostly need the case covered by (5.8).

5.7 Lemma. *Let M be a topological space and Γ a discontinuous group of homeomorphisms of M. Then there is a unique topological space $\Gamma \backslash M$*

(called the quotient of M by Γ) and a continuous Γ-invariant map $p : M \to \Gamma\backslash M$ such that every continuous Γ-invariant map factors through p.

If the action of Γ is free then p is a local homeomorphism.

If M is a complex analytic space and Γ a group of biholomorphisms then $\Gamma\backslash M$ is a complex analytic space and p is holomorphic. ∎

5.8 Example. Let X be a locally contractible topological space and $p : M \to X$ a covering space corresponding to a quotient Γ of $\pi_1(X)$. Γ acts on M by deck transformations and this action is discontinuous. The action is fixed point free and $\Gamma\backslash M = X$. Γ has a compact fundamental domain iff X is compact.

5.9 Notation. For the rest of this section let M be a normal complex space and Γ a group of biholomorphisms acting discontinuously. Let L_M be a line bundle on M with a fixed Γ-action. (The choice of a Γ-action is very important. Different choices lead to very different behavior. See for example chapter 8.)

The simplest example of this situation arises when M is the universal covering space of a complex space X and L_M is the pull back of a line bundle L_X from X.

5.10 Definition. Notation as above. Let f_k be a section of L_M^k. The series

$$P(f_k)(z) = \sum_{\gamma\in\Gamma} f_k(\gamma z)$$

is called the *Poincaré series* of f_k. We call k the *weight* of the Poincaré series. $P(f_k)$ is a Γ-invariant section of L_M^k, if the summation is everywhere convergent.

5.11 Remark. By the above definition, the weight of a Poincaré series depends on the choice of L_M. The most natural choice for L_M is the canonical bundle K_M; thus, by our definition, sections of K_M^k give weight k Poincaré series.

When M is the Siegel upper half space H_n, then $K_M = L_M^{n+1}$ for a unique line bundle L_M with an action of the symplectic group. The classical definition of weight uses L_M as the basic choice.

5.12 Further notation. Let us fix a continuous Γ-invariant measure $d\mu$ on M and a continuous Γ-invariant Hermitian metric $h(\, , \,)$ on L_M.

(5.12.1) Let f be a measurable section of L_M. $\|f(x)\| = h(f(x), f(x))^{1/2}$ is a measurable function on M. We say that f is an L^p section of L_M if

$$(5.12.1.1) \quad \|f\|^p_{M,p} := \int_M \|f(x)\|^p d\mu < \infty.$$

$\|f\|_{M,p}$ is called the L^p-norm of f. This definition depends on the choices of $d\mu$ and h. If Γ has a compact fundamental domain, then any two choices are bounded by constant multiples of each other; thus we can say that f is an L^p section of L_M without specifying $d\mu$ or h.

(5.12.2) The space of holomorphic L^p sections is denoted by $(L^p \cap H)(M, d\mu, L_M, h)$ or simply by $(L^p \cap H)(M, L_M)$ if no confusion is likely.

(5.12.3) The metric h induces a metric h^m on $L_M^{\otimes m}$. It is clear that f is L^p iff $f^{\otimes m}$ is $L^{p/m}$. In particular, if f is L^2 then f^2 is L^1.

5.13 Example. There is one very important special case where we do not have to choose a lifting of the Γ-action, a metric h, or a measure $d\mu$. This happens when M is a manifold, L_M is the canonical bundle K_M, and $p = 2$. One can define a natural L^2-norm on sections of K_M by

$$\|g\|^2_M = c(\dim M) \int_M g \wedge \bar{g} \quad \text{where} \quad c(n) = (-2i)^{-n}(-1)^{\frac{n(n-1)}{2}}.$$

(The strange constant comes from the identity

$$\begin{aligned}(-2i)^{-n}(-1)^{\frac{n(n-1)}{2}} dz_1 \wedge \cdots \wedge dz_n \wedge d\bar{z}_1 \wedge \cdots \wedge d\bar{z}_n \\ = dx_1 \wedge dy_1 \wedge \cdots \wedge dx_n \wedge dy_n.)\end{aligned}$$

I claim that this is equivalent to the L^2-norm defined in (5.12.1.1) if Γ has a compact fundamental domain.

In order to see this we need to compare the $2n$-forms $c(n) g \wedge \bar{g}$ and $\|g\|^2 d\mu$. Pick a point $x \in M$ and local coordinates z_i at x. Let

$$\begin{aligned} g &= g(\mathbf{z}) dz_1 \wedge \cdots \wedge dz_n, \\ h(\mathbf{z}) &= \|dz_1 \wedge \cdots \wedge dz_n\|, \quad \text{and} \\ d\mu &= \mu(\mathbf{z})^2 c(n) dz_1 \wedge \cdots \wedge dz_n \wedge d\bar{z}_1 \wedge \cdots \wedge d\bar{z}_n. \end{aligned}$$

Thus

$$\begin{aligned} g \wedge \bar{g} &= |g(\mathbf{z})|^2 dz_1 \wedge \cdots \wedge dz_n \wedge d\bar{z}_1 \wedge \cdots \wedge d\bar{z}_n, \quad \text{and} \\ \|g\|^2 d\mu &= c(n) h(\mathbf{z})^2 \mu(\mathbf{z})^2 |g(\mathbf{z})|^2 dz_1 \wedge \cdots \wedge dz_n \\ &\quad \wedge d\bar{z}_1 \wedge \cdots \wedge d\bar{z}_n. \end{aligned}$$

Therefore

$$\|g\|^2 d\mu/c(n)g \wedge \bar{g} = h(\mathbf{z})^2\mu(\mathbf{z})^2$$

is a strictly positive Γ-invariant function on M. Thus it is bounded from below and from above if Γ has a compact fundamental domain. ∎

There is one important case when it is easy to construct L^2 sections of the canonical bundle.

5.13.1 LEMMA. *Let Y be a Stein manifold and let $M \subset Y$ be a bounded open set. Then K_M is generated by its L^2 sections.*

Proof. Pick $x \in M$. Choose n holomorphic functions g_i such that the dg_i are independent at x and take $\omega = dg_1 \wedge \cdots \wedge dg_n$. ω is a section of K_Y which is nonzero at x. $M \subset Y$ is realtively compact, thus

$$c(n)\int_M \omega \wedge \bar{\omega} < \infty. \quad ∎$$

5.14 LEMMA. *Notation as in (5.12). Assume $0 < p < \infty$. Assume that f is L^p and let $q \geq p$.*

(5.14.1) The series $\sum_{\gamma\in\Gamma} \|f(\gamma x)\|^q$ is uniformly convergent on compact sets of M.

(5.14.2) If Γ has a compact fundamental domain, then f is L^q. In fact, the inclusion

$$(L^p \cap H)(M, L_M) \to (L^q \cap H)(M, L_M)$$

is continuous for $q \geq p$.

Proof. Let $x \in U \subset M$ be as in (5.5.2). Let $B \subset \mathbb{C}^n$ be the unit ball. Pick a biholomorphism $n : B \hookrightarrow U$ such that $n(0) = x$. We may assume that n extends to a slightly larger ball. Let s be a nowhere zero holomorphic section of $L_M|U$. The choice of s gives an ismorphism $n^*(L|U) \cong \mathcal{O}_B$. By translation we obtain $\gamma \circ n : B \to \gamma U$. Set $\Gamma = \{\gamma_i\}$ and define holomorphic functions on B by $g_i = n^*\gamma_i^* f$. There are constants $c_1, c_2 > 0$ such that

$$c_1^{-1}|n^*h(z)| \leq \|h(n(z))\| \leq c_1|n^*h(z)|, \quad \text{and}$$

$$c_2^{-1}n^*d\mu \leq dm \leq c_2 n^*d\mu,$$

for every $h \in \Gamma(U, L_M|U)$ and every $z \in B$, where dm is the standard Lebesgue measure on B. $|g_\iota(z)|^p$ is plurisubharmonic on B. By the mean value theorem, if $|z| \leq 1/2$ then

$$|g_\iota(z)|^p \leq \frac{2^{2n}}{\operatorname{vol} B} \int_{z+B(1/2)} |g_\iota|^p dm. \tag{5.14.3}$$

Therefore

$$\begin{aligned}
\sum_{\gamma \in \Gamma} \|f(\gamma(n(z)))\|^p &\leq c_1^p \sum_\iota |g_\iota(z)|^p \leq \frac{c_1^p 2^{2n}}{\operatorname{vol} B} \sum_\iota \int_B |g_\iota|^p dm \\
&\leq \frac{c_1^{2p} c_2 2^{2n}}{\operatorname{vol} B} \sum_{\gamma \in \Gamma} \int_{\gamma U} \|f\|^p d\mu \\
&\leq \frac{c_1^{2p} c_2 |\operatorname{Stab}(x)| 2^{2n}}{\operatorname{vol} B} \|f\|_{M,p}^p < \infty.
\end{aligned}$$

Since $q \geq p$ we obtain

$$\sum_{\gamma \in \Gamma} \|f(\gamma(n(z)))\|^q \leq \left(\frac{c_1^{2p} c_2 |\operatorname{Stab}(x)| 2^{2n}}{\operatorname{vol} B} \|f\|_{M,p}^p \right)^{q/p} < \infty.$$

Therefore, if $K \subset M$ is compact then there is a constant $c(K)$ such that

$$\sum_\gamma \|f(\gamma x)\|^q \leq c(K) \|f\|_{M,p}^q, \quad \forall x \in K.$$

If K is also a fundamental domain then

$$\int_M \|f(x)\|^q \leq \int_K \sum_\gamma \|f(\gamma x)\|^q \leq \operatorname{vol}(K) c(K) \|f\|_{M,p}^q. \quad \blacksquare$$

5.15 Corollary. *Notation as above. Let $f \in (L^p \cap H)(M, L_M)$ and g a bounded holomorphic function on M. Let $k \geq p$ be an integer. Then the Poincaré series $P(f^k g)(z) = \sum_\gamma f^k(\gamma z) g(\gamma z)$ is convergent and defines a Γ-invariant holomorphic section of L_M^k.* $\blacksquare$

5.16 Definition. Let M be a complex manifold. We say that (bounded) holomorphic functions *separate points* of M if for every $p \neq q \in M$ there is a (bounded) holomorphic function f on M such that $f(p) \neq 0$ and $f(q) = 0$.

This implies that for any finite collection of points $p_i \in M$ and for any $c_i \in \mathbb{C}$ there is a (bounded) holomorphic function f on M such that $f(p_i) = c_i$ for every i.

5.17 THEOREM. *Notation as above. Assume that Γ operates on M discontinuously. Assume that bounded holomorphic functions separate the points of M. Let $f \in (L^p \cap H)(M, L_M)$ for some $p > 0$. Let $x, y \in M$ be two points such that $\Gamma x \neq \Gamma y$. Assume that f is not identically zero on Γx.*

Then for every $k \gg 1$ which is divisible by $|\operatorname{Stab}(x)|$ there is a bounded holomorphic function g_k on M such that

$$P(f^k g_k)(x) \neq 0 \quad \textit{and} \quad P(f^k g_k)(y) = 0.$$

Proof. $\operatorname{Stab}(x)$ acts on $L_M(x)$ (the fiber of L_M at x). If this action is nontrivial, then every $\operatorname{Stab}(x)$ invariant section on L_M vanishes at x. This is the reason for the assumption that $|\operatorname{Stab}(x)|$ divides k. Set $|\operatorname{Stab}(x)| = d(x)$ and $|\operatorname{Stab}(y)| = d(y)$. Replacing L_M by $L_M^{d(x)}$ we may assume that $\operatorname{Stab}(x)$ acts trivially on $L_M(x)$. Thus we can forget about the divisibility assumption for the rest of the proof.

Let $\Gamma x = \{x_1, x_2, \dots\}$ and $\Gamma y = \{y_1, y_2, \dots\}$. Pick $x_1 \in \Gamma x$ such that $f(x_1) \neq 0$. Dividing f by $\|f(x_1)\|$ we may assume that $\|f(x_1)\| = 1$. Let $c_0 = \max\{\|f(y_i)\|\}$ and $c_1 = 2\max\{1, c_0\}$. By choosing $m \gg 1$ we may assume that

$$\|f(x_i)\| < 1/c_1, \quad \|f(y_i)\| < 1/c_1 \quad \text{for } i > m.$$

Let q_1 be a bounded holomorphic function such that

$$q_1(x_1) = 1 \quad \text{and}$$
$$q_1(x_2) = \dots = q_1(x_m) = q_1(y_1) = \dots = q_1(y_m) = 0.$$

Let c_2 be an upper bound for $|q_1|$ on M and let c_3 be an upper bound for

$$\sum_{i>m} \|f(x_i)\|^p \quad \text{and} \quad \sum_{i>m} \|f(y_i)\|^p.$$

If $k > p + r$ then, since $P(f^k q_1)(x) = d(x) \sum f^k q_1(x_i)$, we obtain

$$\begin{aligned} \big| \|P(f^k q_1)(x)\| - d(x) \big| &\le d(x) c_1^{-r} c_2 c_3 \quad \text{and} \\ \|P(f^k q_1)(y)\| &\le d(y) c_1^{-r} c_2 c_3. \end{aligned} \tag{5.17.1}$$

If $f|\Gamma y$ is identically zero or $\mathrm{Stab}(y_1)$ does not act trivially on $L_M(y_1)$, then $P(f^k q_1)(y) = 0$ and we are done. Otherwise we may assume that $\|f(y_1)\| = c_0$ and $\mathrm{Stab}(y_1)$ acts trivially on $L_M(y_1)$.

Choose q_2 such that

$$q_2(x_1) = q_2(y_1) = 1 \quad \text{and}$$
$$q_2(x_2) = \ldots = q_2(x_m) = q_2(y_2) = \ldots = q_2(y_m) = 0.$$

We may assume that c_1 is an upper bound for $|q_2|$ as well. As above we obtain that

$$(5.17.2)\qquad \begin{aligned} &\big|\|P(f^k q_2)(x)\| - d(x)\big| \leq d(x)c_1^{-r}c_2c_3 \quad \text{and} \\ &\big|\|P(f^k q_2)(y)\| - d(y)c_0^{-r}\big| \leq d(y)c_1^{-r}c_2c_3. \end{aligned}$$

Take

$$g_k = q_1 - \frac{P(f^k q_1)(y)}{P(f^k q_2)(y)} q_2.$$

By construction $P(f^k g_k)(y) = 0$. Since $\|P(f^k q_1)(y)\| / \|P(f^k q_2)(y)\| \to 0$ as $k \to \infty$, we see that $\|P(f^k g_k)(x)\| \to d(x)$ as $k \to \infty$. ∎

Our aim is to prove that under certain conditions the Poincaré series of large weight separate any two points. It is convenient to set up a more general framework.

5.18 Definition. Let X be a complex space and L a line bundle on X. Let

$$R = \sum R_k \subset \sum_{k=0}^{\infty} H^0(X, L^k)$$

be a graded subring such that every R_k is finite dimensional. Any choice of a basis in R_k determines a meromorphic map $\phi_{R_k} : X \dashrightarrow \mathbb{P}$ to a projective space of dimension $\dim R_k - 1$.

5.19 Definition. Let X be a complex space and L a line bundle on X. Let

$$R = \sum R_k \subset \sum_{k=0}^{\infty} H^0(X, L^k)$$

be a graded subring such that every R_k is finite dimensional.

(5.19.1) Let $\mathrm{Bs}\,|R_k| \subset X$ be the closed subset where all elements of R_k vanish (the *base locus* of the sections in R_k).

(5.19.2) Let $\phi_{R_k} : X \setminus \mathrm{Bs}\,|R_k| \to \mathbb{P}$ be the morphism to a projective space of dimension $\dim R_k - 1$ given by R_k.

(5.19.3) Let

$$F_{R_k} = (\mathrm{Bs}\,|R_k| \times X) \cup (X \times \mathrm{Bs}\,|R_k|) \\ \cup \{(x_1, x_2) | \phi_{R_k}(x_1) = \phi_{R_k}(x_2)\} \subset X \times X.$$

F_{R_k} measures the fibers of ϕ_{R_k}.

5.20 Lemma. *Notation as above.*

(5.20.1) F_{R_k} = diagonal of $X \times X$ iff ϕ_{R_k} is everywhere defined and is one-to-one on closed points.

(5.20.2) The diagonal of $X \times X$ is an irreducible component of F_{R_k} iff ϕ_{R_k} is generically finite.

(5.20.3) If $k|m$ then $F_{R_m} \subset F_{R_k}$.

(5.20.4) Assume that X is proper (or X is algebraic and we consider the algebraic $H^0(X, L^k)$). Then there is an $N = N(X, R)$ such that $F_{R_{kN}} = F_{R_N}$ for every $k \geq 1$. Denote F_{R_N} by F_R. ■

5.21 Definition. Notation as in (5.9–10). Let

$$P(M, \Gamma, L_M) = \sum P_k \subset \sum H^0(M, L_M^k)$$

be the subalgebra generated by all Poincaré series $P(f)$ where f is a section of L_M^m for some m such that the corresponding Poincaré series is convergent. A general element of P_k is a linear combination of products of Poincaré series such that the sum of weights is k.

We call $P(M, \Gamma, L_M)$ the algebra of Poincaré series on M or on $\Gamma\backslash M$.

5.22 Theorem. *(Poincaré) Let $M \subset Y$ be a bounded open subset of a Stein manifold Y (i.e., the closure of M in Y is compact). Let Γ be a discrete group acting freely and discontinuously on M. Assume that $X = \Gamma\backslash M$ is compact. Then K_X is ample.*

More precisely, the degree k graded piece of the algebra of Poincaré series on X gives an embedding of X for $k \gg 1$.

Proof. We only prove that for $k \gg 1$, the degree k piece of the algebra of Poincaré series on X defines a morphism that is a homeomorphism onto its image. In order to get an embedding one needs to prove an analog of (5.17) about separating tangent vectors. The proof is very similar to (5.17).

By (5.20) it is sufficient to prove that if $x_1 \neq x_2 \in X$, then there is a Poincaré series that is nonzero at x_1 and zero at x_2.

Let $p, q \in M$. Since Y is Stein, there is a holomorphic function f on Y such that $f(p) = 0$ and $f(q) \neq 0$. Thus $f|M$ is a bounded function separating p and q. By (5.13.1) K_M is generated by its L^2 sections.

From (5.17) we obtain a Poincaré series which separates x_1 and x_2. ∎

5.23 Complement. Theorem (5.22) can be generalized to certain cases when γ has fixed points and X is not compact.

It happens frequently that $\Gamma\backslash M$ is not compact, but has an algebraic structure and every holomorphic section of K_X^m is algebraic. This is the case for the standard action of the integral symplectic group on the Siegel upper half space. (Koecher principle; see, e.g., [Klingen90, p. 45]). Our proof applies verbatim in this case.

If the fixed point set of every $1 \neq \gamma \in \Gamma$ has codimension at least 2 in M, then K_M descends to the $\mathbb{Q}$-line bundle K_X and the Poincaré series give sections of powers of K_X. One should, however, keep in mind that these sections need not be pluricanonical forms on a desingularization of X. (Their pullback to a desingularization can pick up poles along the exceptional divisors.) This situation can be completely analyzed by studying the action of the stabilizer subgroups near the fixed points. (See, e.g., [Reid80].)

If some elements of Γ have fixed points in codimension one, then K_M descends to a $\mathbb{Q}$-line bundle of the form

$$K_X + \sum \left(1 - \frac{1}{e(D)}\right) D,$$

where $D \subset X$ is a divisor and $e(D)$ is the ramification index of $M \to X$ above D. Alternatively, we may always assume that only the identity acts trivially on M and then $e(D)$ is the order of the subgroup of Γ leaving the preimage of D pointwise fixed.

With these changes in mind, the above proof applies to get an ampleness theorem for these more general quotients.

5.24 Remark. From the point of view of the general philosophy of higher-dimensional geometry it would be very natural to consider the case when M is (mildly) singular. The easiest would be to allow isolated singularities. Γ acts on the singular set, thus if M is not smooth then we would get a series of isolated singularities converging to the boundary of M in Y. This is impossible. I do not know a single example of a Stein space Y and a singular bounded open set $M \subset Y$ such that $\mathrm{Aut}(M)$ has a compact fundamental domain.

CHAPTER 6

The Method of Atiyah

Let X be a compact Kähler manifold and E a vector bundle on X with a Hermitian metric. By Hodge theory, the cohomology $H^i(X, E)$ can be represented by E-valued harmonic $(0, i)$-forms. Let $\tilde{X}$ be the universal cover of X. We can pull back the Kähler form, E and its metric to $\tilde{X}$. The definitions of Hodge theory being local, harmonic forms make sense on $\tilde{X}$. Thus we conclude that $H^i(X, E)$ can be represented by $\tilde{E}$-valued harmonic $(0, i)$-forms on $\tilde{X}$ which are invariant under $\pi_1(X)$.

Unfortunately $\pi_1(X)$-invariant forms are not in any L^p space ($p < \infty$); thus it is very hard to do analysis with them. It is much more useful to have a result that connects the cohomology of E with harmonic L^2 forms on $\tilde{X}$. Already for finite covers this works well only on the level of the Euler characteristic. The general result is given by the L^2 index theorem of Atiyah, which we now recall. In order to stay in a framework more familiar to algebraic geometers, we consider only the case of the cohomology of a vector bundle instead of a general elliptic differential operator. It should be emphasized, however, that in order to prove the L^2-index theorem it is more natural, in fact essential, to use the methods of the Atiyah-Singer index theorem.

6.1 Hirzebruch-Riemann-Roch theorem. Let Z be a smooth projective variety of dimension n over $\mathbb{C}$ and E a vector bundle on Z. Pick Hermitian metrics g on T_Z and h on E. These give differential forms $c_i(Z, g)$ (resp. $c_i(E, h)$) which represent the ith Chern class of Z (resp. E).

6.1.1 THEOREM. *Notation as above. There is a polynomial $P(x_1, \dots, x_n, y_1, \dots, y_n)$ depending only on n such that P is homogeneous of degree $2n$ if we set $\deg x_i = \deg y_i = 2i$, and*

$$\chi(Z, E) = \int_Z P(c_1(Z, g), \dots, c_n(Z, g), c_1(E, h), \dots, c_n(E, h)). \quad \blacksquare$$

The precise version [Hirzebruch66] actually identifies the polynomial P. We do not need this for now.

Let $p : \tilde{Z} \to Z$ be a finite étale Galois cover and let us pull everything back to $\tilde{Z}$. It is clear that $p^*(c_i(Z, g)) = c_i(\tilde{Z}, \tilde{g})$, hence the right-hand side of (6.1.1) is multiplicative in étale covers. Thus we obtain the following.

6.1.2 COROLLARY. *Let $p : \tilde{Z} \to Z$ be a finite étale Galois cover and $\tilde{E} = p^*E$. Then*

$$\chi(\tilde{Z}, \tilde{E}) = \deg(\tilde{Z}/Z)\chi(Z, E). \quad \blacksquare$$

6.2 L^2 Index Theorem. We would like to get a statement similar to (6.1.2) for infinite covers where we want to consider the L^2-cohomology of $\tilde{E}$. We are faced with two fundamental problems:

(6.2.1.1) The various definitions of cohomology groups are inequivalent for noncompact manifolds.

(6.2.1.2) The resulting cohomology groups are usually infinite dimensional, thus the alternating sum of their dimensions does not make sense.

The answer to the first problem is clear if one looks at it from the point of view of elliptic operators.

6.2.2 Definition. Let M be a complex manifold (compact or not) and F a vector bundle on M. Fix a Kähler metric ω on M and a Hermitian metric h on F. Let $\mathcal{H}^{(i,j)}_{(2)}(M, \omega, F, h)$ be the space of F-valued, L^2, harmonic (i, j) forms on M.

If the choice of ω and h is clear from the context, we drop them from the notation. It should be emphasized that in the noncompact case the definition depends on the choice of the metrics.

As usual, $\mathcal{H}^{(0,j)}_{(2)}(M, F)$ is called the jth *L^2-cohomology group* of F on M.

We also use the notation $H^0_{(2)}(M, F)$ to denote the space of holomorphic L^2-sections. Clearly,

$$H^0_{(2)}(M, F) = (L^2 \cap H)(M, F) = \mathcal{H}^{(0,0)}_{(2)}(M, F).$$

The proliferation of notation is unfortunate but it reflects the state of the literature.

In order to solve the second problem we need a notion of dimension that yields a finite number even for some infinite dimensional spaces.

Such a definition is provided by the theory of von Neumann algebras. One unusual feature is that the dimension is a real valued function. Thus it is not a priori clear that the right-hand side of (6.3) is an integer (or even a rational number).

Much of the necessary background material is explained very well in [Atiyah76]. See also [Arveson76, chap. 1] for a proof of basic properties, and [Dixmier81] for a more complete treatment. In the case that is of interest to us, it is possible to write down the definition of the dimension quite explicitly. This is done after (6.5).

Accepting the definition on faith for a moment, we can formulate the L^2 index theorem. We make a twist and consider $K_X \otimes E$ instead of E. $K_X \otimes E$ valued $(0, i)$ forms can be identified with E valued (n, i) forms. Thus we obtain the following special case of the L^2 index theorem of [Atiyah76].

6.3 THEOREM. *Let X be a compact Kähler manifold of dimension n with fundamental group Γ, and E a holomorphic vector bundle with a Hermitian metric. Let $\mathcal{H}_{(2)}^{(n,i)}(\tilde{X}, \tilde{E})$ be the space of $\tilde{E}$-valued, L^2, harmonic (n, i) forms on $\tilde{X}$ (where we use the pullback metrics on $\tilde{X}$ and on $\tilde{E}$). Then*

$$\chi(X, K_X \otimes E) = \sum_i (-1)^i \dim_\Gamma \mathcal{H}_{(2)}^{(n,i)}(\tilde{X}, \tilde{E}). \quad \blacksquare$$

6.3.1 Remark. It is frequently very desirable to have a version of (6.3) when X is not compact, for instance when X is a Zariski open subset of a projective variety $\bar{X}$. Under reasonable conditions (see, e.g., [Cheeger-Gromov85]) one obtains a formula

$$\text{(6.3.1.1)} \quad \int_X P(c_j(X, g), c_j(E, h)) = \sum_i (-1)^i \dim_\Gamma \mathcal{H}_{(2)}^{(n,i)}(\tilde{X}, \tilde{g}, \tilde{E}, \tilde{h}).$$

Unfortunately, the intergal on the left-hand side does not compute the L^2-cohomology of X. One needs some correction terms coming from the boundary of X.

By the philosophy of [CGM82], the L^2-cohomology of E on X should compute a suitable intersection cohomology on $\bar{X}$. Therefore we cannot expect a multiplicative behavior under étale covers of X itself.

We can view (6.3) as a theorem about the existence of automorphic forms. If we know enough about harmonic L^2-froms with values in $\tilde{E}$, then we can get some information about the cohomologies of E. On the one hand, the result is much more precise since it provides information about E itself (insted of high tensor powers). On the other hand, the answer involves higher cohomologies, which makes it more difficult to interpret in some situations.

The result becomes especially strong if the higher cohomologies vanish. We formulate here a clasical case, which is further generalized in chapter 11.

6.4 THEOREM. *Notation as in (6.3). Let L be an ample line bundle on X. Then*

$$h^0(X, K_X \otimes L) = \dim_\Gamma H^0_{(2)}(\tilde{X}, K_{\tilde{X}} \otimes \tilde{L}).$$

Proof. By the Kodaira vanishing theorem (9.1), $h^0(X, K_X \otimes L) = \chi(X, K_X \otimes L)$. Furthermore,

$$H^0_{(2)}(\tilde{X}, K_{\tilde{X}} \otimes \tilde{L}) = \mathcal{H}^{(n,0)}_{(2)}(\tilde{X}, \tilde{L}).$$

Thus it is sufficient to prove that

$$\text{(6.4.1)} \quad \mathcal{H}^{(n,i)}_{(2)}(\tilde{X}, \tilde{L}) = 0 \quad \text{for } i > 0.$$

This is a result of [Andreotti-Vesentini65]. We return to a more general form in chapter 11. ∎

The method of Poincaré shows that there are plenty of automorphic forms of sufficiently large weight. Unfortunately, the method does not seem to give information about forms of low weight. The index theorem shows that such forms exist.

6.5 COROLLARY. *Let $M \subset Y$ be a bounded open subset of a Stein manifold Y and Γ a discrete group acting freely and discontinuously on M. Assume that $X = \Gamma\backslash M$ is compact. Then*
(6.5.1) $h^0(X, K_X^m) \geq 1$ for $m \geq 2$, and
(6.5.2) $h^0(X, K_X^m) \geq 2$ for $m \geq 4$.

Proof. We have seen in the proof of (5.22) that K_M is generated by its holomorphic L^2 sections. In particular, K_M^m has nonzero holomorphic $L^{2/m}$ sections. By (5.14.2) these are also L^2, thus by (6.10.4)

$$\dim_\Gamma \mathcal{H}^{(n,0)}_{(2)}(M, K_M^m) > 0.$$

Thus (6.4) implies the first part.

Let $s \in H^0(X, K_X^2)$ be a nonzero section and $\tilde{s}$ its pullback on M. Consider the Γ-equivariant multiplication map

$$m(\tilde{s}) : H^0_{(2)}(M, K_M^m) \longrightarrow H^0_{(2)}(M, K_M^{m+2}).$$

It is easy to see that $m(\tilde{s})$ is injective with closed image (11.2.3). The zero set of $\tilde{s}$ is not empty since K_X^2 is not the trivial line bundle. Thus $\operatorname{im} m(\tilde{s})$ does not generate $H^0_{(2)}(M, K_M^{m+2})$. If $m \geq 2$ then by (6.4) and (6.11.1)

$$\begin{aligned} h^0(X, K_X^{m+2}) &= \dim_\Gamma H^0_{(2)}(M, K_M^{m+2}) \\ &> \dim_\Gamma H^0_{(2)}(M, K_M^m) = h^0(X, K_X^m). \quad \blacksquare \end{aligned}$$

6.5.3 Remarks. The example of [Mumford79] gives a two-dimensional ball quotient with $h^0(X, K_X) = 0$. Thus (6.5.1) is optimal in this sense.

If X is an arbitrary smooth compact surface of general type, then $h^0(X, K_X^2) \geq 2$. It is possible that (6.5.2) holds for $m \geq 2$.

Moreover, in all the examples I know, K_X^2 is generated by global sections and K_X^3 is very ample. Examples in dimensions of at least 3 are very hard to describe, so this may be a low-dimensional phenomenon.

It is now time to return to the basics and work through an elementary exposition of traces on a special class of von Neumann algebras. These definitions (and a few results) are needed for the statement and some applications of the L^2 index theorem.

6.6 Definition. For a Hilbert space H let $B(H)$ denote the algebra of bounded linear operators $H \to H$.

Our aim is to define the trace of an operator in $B(H)$ and in some subalgebras of $B(H)$. This is done in four steps.

6.7 **Step 1.** Let H be a finite dimensional Hilbert space with an orthonormal basis $\{e_i\}$. For $A \in B(H)$ let $\operatorname{Tr}(A) = \sum_i (Ae_i, e_i)$. It is easy to see that this is independent of the choice of the basis, and if A is given by a matrix $A = (a_{ij})$ then $\operatorname{Tr}(A) = \sum a_{ii}$.

6.7.1 Definition. Let H be a Hilbert space.

(6.7.1.1) We call an operator $A \in B(H)$ positive if $(Ah, h) \geq 0$ for every $h \in H$. It is not difficult to see that a positive operator is selfadjoint [Conway90, II.2.12] but we can just assume that this is part of the definition.

(6.7.1.2) $B = (b_{ij})$ is called a Hilbert-Schmidt operator if $||B|| = \sum |b_{ij}|^2 < \infty$. These are the operators where we have really no convergence problems and it is easy to check that $||B||$ is independent of the choice of basis.

6.8 **Step 2.** Let H be an infinite dimensional Hilbert space with an orthonormal basis $\{e_i\}$. We want to follow the definition of trace given for finite dimensional Hilbert spaces. The sum $\sum(Ae_i, e_i)$ need not converge in general. Even if it does, it may depend on the choice of basis. There are two important classes of operators where one can define trace with good properties.

(6.8.1) If A is positive then let $\mathrm{Tr}(A) = \sum_i(Ae_i, e_i)$. This version of trace is defined on the set of positive operators and takes values in $[0, \infty]$. If $B = (b_{ij})$ is a bounded operator then $A = BB^*$ is positive and $\mathrm{Tr}(A) = \sum |b_{ij}|^2$. In particular $\mathrm{Tr}(BB^*) = \mathrm{Tr}(B^*B)$.

(6.8.2) If $A = BC$ is the product of two Hilbert-Schmidt operators, then set $\mathrm{Tr}(A) = \sum_i(Ae_i, e_i) = \sum_i(Ce_i, B^*e_i)$.

It is not hard to see that these definitions coincide if they are both defined on A. The definitions are independent of the choice of the basis. This can be expressed in a more invariant way: if $U \in B(H)$ is unitary then $\mathrm{Tr}(U^{-1}AU) = \mathrm{Tr}(A)$.

Note that $\mathrm{Tr}(AB) = \mathrm{Tr}(BA)$ holds if A, B are Hilbert-Schmidt operators, but not in general (even though both sides may be defined).

6.8.3 Lemma. *Let A be positive. Then* $\mathrm{Tr}(A) = 0 \Leftrightarrow A = 0$.

Proof. By positivity, $0 \leq (A(se_i + te_j), se_i + te_j)$ for every s, t. Viewing this as a Hermitian form in s, t shows that $|a_{ij}|^2 \leq |a_{ii}||a_{jj}|$. $\mathrm{Tr}(A) = 0$ iff $a_{ii} = 0$ for every i. ∎

(6.8.4) Let $P \in B(H)$ be a projection onto a subspace $H' \subset H$. Choose an ONB (e_i, f_j) such that e_i is a basis of H' and f_j is a basis for the orthogonal complement. Then $\mathrm{Tr}(P) = \sum(e_i, e_i) = \dim H'$.

6.9 **Step 3.** Let Γ be a discrete group and $H = L^2(\Gamma)$. H has a basis consisting of functions $\delta_g : g \in \Gamma$ defined by

$$\delta_g(g') = \begin{cases} 1 & \text{if } g = g', \text{ and} \\ 0 & \text{if } g \neq g'. \end{cases}$$

Let l_g (resp. r_g) denote the left (resp. right) translation by g: $l_g(\delta_{g'}) = \delta_{gg'}$ and $r_g(\delta_{g'}) = \delta_{g'g^{-1}}$.

6.9.1 Definition. Let $A(\Gamma) \subset B(L^2(\Gamma))$ be the algebra of all operators that commute with all left translations.

Let $A \in A(\Gamma)$ be given by the matrix $(a_{g,h} : g, h \in \Gamma)$. Since A commutes with left translations,

$$A(\delta_g) = A(l_g(\delta_1)) = l_g(A(\delta_1)) = l_g\left(\sum_h a_{1,h}\delta_h\right) = \sum_h a_{1,g^{-1}h}\delta_h.$$

Thus $a_{g,h} = a_{1,g^{-1}h}$ and there is a function $a : \Gamma \to \mathbb{C}$ such that $a_{g,h} = a(g^{-1}h)$. This implies (ignoring convergence a little) that

$$A = \sum a(h^{-1})r_h,$$

i.e., $A(\Gamma)$ is generated by right translations.

We would like to define a trace on $A(\Gamma)$. Following the previous definitions would give $\sum a_{gg} = \sum a(g^{-1}g) = |\Gamma|a(1)$, which is always infinite unless Γ is finite or $A = 0$. We can, however, renormalize and declare that

$$(6.9.1.1) \quad \mathrm{Tr}(A) := (A\delta_1, \delta_1) = a(1).$$

Tr is defined for every $A \in A(\Gamma)$, and it is additive: $\mathrm{Tr}(A + B) = \mathrm{Tr}(A) + \mathrm{Tr}(B)$. There is another important property of the ordinary trace. if U is unitary then $\mathrm{Tr}(U^{-1}AU) = \mathrm{Tr}(A)$. Something much stronger holds in this case.

6.9.2 LEMMA. *Let* $A, B \in A(\Gamma)$. *Then* $\mathrm{Tr}(AB) = \mathrm{Tr}(BA)$.

Proof. Let $A = (a_{g,h} = a(g^{-1}h))$ and $B = (b_{g,h} = b(g^{-1}h))$. $(A\delta_1, A\delta_1) = \sum_g |a(g)|^2$ is finite. Direct computation gives that

$$(AB)_{g,h} = \sum_f a(g^{-1}f)b(f^{-1}h),$$

and the sum is covergent. Thus

$$\mathrm{Tr}(AB) = (AB)_{1,1} = \sum_f a(f)b(f^{-1}) = \sum_f b(f)a(f^{-1})$$
$$= (BA)_{1,1} = \mathrm{Tr}(BA). \quad \blacksquare$$

(6.9.3) Let $P(H')$ be the orthogonal projection onto $H' \subset L^2(\Gamma)$. If $P(H')$ commutes with left translations then H' is left invariant under Γ. Conversely, if H' is left invariant under Γ then the corresponding orthogonal projection $P(H')$ is in $A(\Gamma)$. We define

$$\dim_\Gamma H' := \mathrm{Tr}\, P(H').$$

Choose an ONB $\{e_i, f_j\}$ such that e_i is a basis of H' and f_j is a basis for the orthogonal complement. Let $e_i = \sum e_i(g)\delta_g$. Then

$$\delta_1 = \sum(\delta_1, e_i)e_i + \sum(\delta_1, f_j)f_j = \sum \bar{e}_i(1)e_i + \sum(\delta_1, f_j)f_j.$$

Thus

(6.9.4) $\quad \dim_\Gamma H' = (P\delta_1, \delta_1) = (\sum \bar{e}_\iota(1)e_\iota, \delta_1) = \sum |e_\iota(1)|^2.$

6.9.5 LEMMA. *Let $H' \subset L^2(\Gamma)$ be a closed subspace invariant under left translations.*

(6.9.5.1) $0 \le \dim_\Gamma H' \le 1$.

(6.9.5.2) $\dim_\Gamma H' = 0$ *iff* $H' = 0$ *and* $\dim_\Gamma H' = 1$ *iff* $H' = L^2(\Gamma)$.

(6.9.5.3) If $H' \subset H''$ *then* $\dim_\Gamma H' \le \dim_\Gamma H''$ *with equality holding iff* $H' = H''$.

Proof. Let $P = (p(g^{-1}h))$ be the projection to H'. Assume that $\mathrm{Tr}(P) = 0$, i.e., $p(1) = 0$. Then $p_{gg} = p(1) = 0$; thus by (6.8.3) $P = 0$. Hence $\dim_\Gamma H' = 0$ iff $H' = 0$. The rest of (6.9.5.1) and (6.9.5.2) are clearly consequences of (6.9.5.3).

Let H^t be the orthogonal complement of H' in H''. Then $P(H'') = P(H') + P(H^t)$ and $P(H')P(H^t) = 0$. Thus

$$\begin{aligned}(P(H'')\delta_1, \delta_1) &= (P(H'')^2\delta_1, \delta_1) \\ &= (P(H'')\delta_1, P(H'')\delta_1) \\ &= (P(H')\delta_1, P(H')\delta_1) + (P(H^t)\delta_1, P(H^t)\delta_1) \\ &= (P(H')\delta_1, \delta_1) + (P(H^t)\delta_1, \delta_1). \quad \blacksquare\end{aligned}$$

6.9.6 Example. The Fourier series expansion

$$f \mapsto \sum_{n\in\mathbb{Z}} e^{2\pi i n t} \int_0^1 f(x)e^{-2\pi i n x}\,dx$$

gives an isomorphism $L^2(\mathbb{R}/\mathbb{Z}) \cong L^2(\mathbb{Z})$. Translation in $L^2(\mathbb{Z})$ corresponds to multiplication by $e^{2\pi i x}$ in $L^2(\mathbb{R}/\mathbb{Z})$, and the basis $\{\delta_n\}$ corresponds to the basis $\{e^{-2\pi i n x}\}$.

Let $U \subset [0, 1]$ be a measurable set and $H_U \subset L^2(\mathbb{R}/\mathbb{Z})$ the subspace of those functions that vanish on U. H_U is invariant under multiplication by $e^{2\pi i x}$. The orthogonal projection to H_U is given by $P(H_U)f = \chi_U \cdot f$ where χ_U is the characteristic function of U. By (6.9.1.1) (we write $\mathbb{Z}$ additively, hence 0 denotes the identity),

$$\dim_{\mathbb{Z}} H_U = (P(H_U)\delta_0, \delta_0) = (\chi_U \cdot 1, 1) = \text{measure}(U).$$

6.10 **Step 4.** Let M be a differentiable manifold with a volume form dV. Let E be a complex vector bundle with a Hermitian metric h (assumed to be measurable only). Let Γ be a discrete group acting on M

such that the action lifts to an action on E. We further assume that dV and h are Γ-equivariant.

Let $L^2(M, E)$ be the space of sections $s : M \to E$ such that $h(s, \bar{s})$ is measurable and $\int_M h(s, \bar{s}) dV < \infty$. This is a Hilbert space, called the space of L^2 sections of E. Clearly Γ acts on $L^2(M, E)$.

Assume that there is an open fundamental domain $M_0 \subset M$ for the Γ action such that its boundary has measure zero. By restriction we obtain E_0, dV and h. It is easy to see that

$$L^2(M, E) \cong L^2(\Gamma) \hat{\otimes} L^2(M_0, E_0),$$

where the isomorphism is given as follows: $L^2(M_0, E_0)$ can be identified with those sections of E which are zero outside M_0. For any $g \in \Gamma$ we can identify $\mathbb{C}\delta_g \otimes L^2(M_0, E_0)$ with those sections of E which are zero outside gM_0. (Here $\hat{\otimes}$ is the tensor product in the Hilbert space sense.)

Let $A(X, E) \subset B(L^2(X, E))$ be those bounded operators that commute with the Γ-action. As in (6.9.1) it is easy to see that every $A \in A(X, E)$ can be written as $A = (a_{g,h} = a(g^{-1}h))$ where $a(f) \in B(L^2(M_0, E_0))$. If the trace of $a(1)$ is defined, then set

(6.10.1) $\quad \operatorname{Tr}_\Gamma(A) := \operatorname{Tr}(a(1)).$

Let $H' \subset L^2(M, E)$ be a closed subspace invariant under Γ and $P(H')$ the corresponding orthogonal projection. $P(H') \in A(X, E)$ and set

(6.10.2) $\quad \dim_\Gamma H' := \operatorname{Tr}_\Gamma(P(H')).$

Let $\{s_\iota\} \subset H'$ be an ONB. As in (6.8.4) and (6.9.4) we obtain the formula

(6.10.3) $\quad \dim_\Gamma H' = \sum_\iota \int_{M_0} h(s_\iota, \bar{s}_\iota) dV.$

As in (6.9.5) we obtain the following elementary properties.

6.10.4 LEMMA. *Let $H' \subset L^2(M, E)$ be a closed subspace invariant under left translations.*

(6.10.4.1) $0 \leq \dim_\Gamma H' \leq +\infty$.

(6.10.4.2) $\dim_\Gamma H' = 0$ *iff* $H' = 0$.

(6.10.4.3) If $H' \subset H''$ *then* $\dim_\Gamma H' \leq \dim_\Gamma H''$. *If* $\dim_\Gamma H' < +\infty$ *then equality holds iff* $H' = H''$. ∎

For later applications we need some easy properties of the dimension function.

6.11 LEMMA. *Notation as in step 4. Let $A \in A(\Gamma)$. Then*

$$\dim_\Gamma(\ker A)^\perp = \dim_\Gamma(\overline{\text{range}\, A}).$$

Proof. The proof has three steps. First we claim that there is an operator $W \in A(\Gamma)$ such that $\ker A = \ker W$ and $W : (\ker A)^\perp \to \overline{\text{range}\, A}$ is an isometry. W is given by the polar decomposition for which we refer to [Conway90, VIII.3.11].

Easy computation gives that WW^* is the orthogonal projection to $\overline{\text{range}\, A}$ and W^*W is the orthogonal projection to $(\ker A)^\perp$. Thus we need to prove that $\text{Tr}(WW^*) = \text{Tr}(W^*W)$.

This holds for any operator B. Write $B = (b_{g,h} = b(g^{-1}h))$. Then $(BB^*)_{1,1} = \sum_h b(h)b^*(h)$; thus

$$\begin{aligned}\text{Tr}_\Gamma(BB^*) &= \sum_h \text{Tr}(b(h)b^*(h)) = \sum_h \text{Tr}(b^*(h)b(h)) \\ &= \text{Tr}_\Gamma(B^*B). \quad \blacksquare\end{aligned}$$

We need the following special case.

6.11.1 COROLLARY. *Notation as in step 4. Let L be a Γ-equivariant line bundle on M and s a Γ-invariant global section of L such that $h(s, \bar{s})$ is bounded. s defines a Γ-equivariant bounded operator $s : L^2(X, E) \to L^2(X, L \otimes E)$. Let $H' \subset L^2(X, E)$ be a Γ-invariant closed subspace and assume that $s|H' : H' \to \overline{s(H')}$ is injective.*

Then $\dim_\Gamma H' = \dim_\Gamma \overline{s(H')}$.

Proof. Let $r = \text{rank}\, E$. We can choose isomorphisms

$$L^2(M_0, E_0) \cong rL^2(M_0) \quad \text{and} \quad L^2(M_0, L_0 \otimes E_0) \cong rL^2(M_0).$$

Let $s_0 = s|M_0$. Multiplication by s becomes a bounded operator

$$L^2(\Gamma)\hat{\otimes} rL^2(M_0) \xrightarrow{id \otimes s_0} L^2(\Gamma)\hat{\otimes} rL^2(M_0).$$

Let $P(H')$ be the projection to H'. Then

$$\begin{aligned}&\ker((id \otimes s_0) \circ P(H'))^\perp = H' \quad \text{and} \\ &\text{range}((id \otimes s_0) \circ P(H')) = s(H').\end{aligned}$$

Thus (6.11) implies (6.11.1). $\blacksquare$

CHAPTER 7

Surjectivity of the Poincaré Map

We keep the notation of chapter 5. Given a complex manifold X and a vector bundle E, we take $\tilde{X}, \tilde{E}$ as before. Let

$$P : (L^1 \cap H)(\tilde{X}, \tilde{E}) \to H^0(X, E)$$

be the Poincaré map (5.10). Our aim is to try to see if P is surjective.

Before we consider the surjectivity of the Poincaré series, we define Bergman kernels in a general setting.

7.1 Bergman kernel.

(7.1.1) Let M be a complex manifold and E a vector bundle on M. We want to define an inner product on sections of E.

The simple way is to consider a volume form dm on M and a Hermitian metric $h(\, , \,)$ on E. On measurable sections of E we obtain an inner product given by

$$(7.1.1.1) \quad f \cdot g = \int_M h(f, g) dm.$$

A more refined version of the same construction is the following.

$E \otimes \bar{E}$ has a natural conjugation given by $\tau : (e \otimes e') \to (e' \otimes e)$. Let $\mathrm{Re}(E \otimes \bar{E})$ denote the real subvector bundle fixed by τ. A Hermitian metric on E is a vector bundle map $E \otimes \bar{E} \to \mathbb{C}_M$, which is real in the sense that $\mathrm{Re}(E \otimes \bar{E})$ is mapped to $\mathbb{R}_M \subset \mathbb{C}_M$.

Let K_M be the canonical bundle and $\Omega_M^{n,n} \cong K_M \otimes \bar{K}_M$ the line bundle of (n, n)-forms. $\Omega_M^{n,n}$ has a natural real structure. In local coordinates $(z_1, \ldots, z_n)$,

$$(\sqrt{-1})^n dz_1 \wedge d\bar{z}_1 \wedge \cdots \wedge dz_n \wedge d\bar{z}_n$$

is declared to be real. The sections of $\mathrm{Re}(\Omega_M^{n,n})$ are exactly the volume forms on M.

With this in mind, we can view $h(\ ,\ )dm$ as a single object:

(7.1.1.2) $H : E \otimes \bar{E} \to \Omega_M^{n,n}$,

where H is a homomorphism of C^∞-vector bundles which is real in the sense that $\mathrm{Re}(E \otimes \bar{E})$ is mapped to $\mathrm{Re}(\Omega_M^{n,n})$.

If dm is a nowhere zero volume form, then set $h = H/dm$ to see that these two definitions are indeed equivalent.

For $E = K_M$ we have a natural isomorphism $K_M \otimes \bar{K}_M \cong \Omega_M^{n,n}$, which corresponds to the natural inner product

(7.1.1.3) $\omega_1 \cdot \bar{\omega}_2 = \int_M \omega_1 \wedge \bar{\omega}_2$.

(7.1.2) By the maximum principle, $(L^2 \cap H)(M, E)$ is a closed subspace of $L^2(M, E)$. Let

(7.1.2.1) $T_E : L^2(M, E) \to (L^2 \cap H)(M, E)$

be the orthogonal projection. Let $\{u_\iota\} \subset L^2(M, E)$ be an orthonormal basis. We can write

(7.1.2.2) $T_E(f) = \sum_\iota (f \cdot \bar{u}_\iota) u_\iota$.

As usual, the projection T_E can also be represented as an integral operator on a product space. To this end consider $M \times M$. We write a point on $M \times M$ as a pair (z, w). Consider the vector bundle $\pi_1^* E \otimes \pi_2^* \bar{E}$ where π_ι are the coordinate projections of $M \times M$. Set

(7.1.2.3) $K_{M,E}(z, w) = \sum_\iota \pi_1^* u_\iota(z) \otimes \pi_2^* \bar{u}_\iota(w)$.

7.2 Lemma-Definition. *(7.2.1) The above sum converges pointwise. On any horizontal or vertical section of $M \times M$ it also converges in the L^2-norm.*

(7.2.2) The resulting section $K_{M,E}(z, w)$ is called the Bergman kernel associated to M, E, H.

(7.2.3) $K_{M,E}(z, w)$ is holomorphic in z and antiholomorphic in w.

(7.2.4) $K_{M,E}(z, w)$ and $K_{M,E}(w, z)$ is are conjugates of each other.

Proof. Fix $w_0 \in M$. At least formally, $K_{M,E}(z, w_0) = T_E(\delta_{w_0})$, where δ_{w_0} is the Dirac delta function centered at w_0. By the maximum principle, the L^2-norm of $K_{M,E}(z, w_0)$ can be estimated from above by the L^2-norm of $T_E(\chi(B(w_0, \epsilon)))$ where $\chi(B(w_0, \epsilon))$ is the characteristic function of a small ball around w_0. Thus the sum in (7.1.2.3) converges in the L^2-norm on any $M \times \{w_0\}$. Moreover, the above argument shows that the L^2-norm of $K_{M,E}|M \times \{w_0\}$ is locally bounded as a function of w_0.

The summands are holomorphic in z and antiholomorphic in w. Therefore the sum converges pointwise to a section of $\pi_1^* E \otimes \pi_2^* \bar{E}$ which is holomorphic in z and antiholomorphic in w. ∎

7.2.5 Complement. We can formally write

$$(7.2.5.1) \quad T_E(f)(z) = \int_M K_{M,E}(z, w) f(w) dm(w).$$

This integral is to be understood as follows. For fixed z_0 we can view $K(z_0, w)$ as a section of $\bar{E}$ on M. Set

$$(7.2.5.2) \quad \int_M K_{M,E}(z_0, w) f(w) dm(w) := \int_M H(f(w), K_{M,E}(z_0, w)).$$

7.3 Remark. In the above situation, E has $\operatorname{rank} E$ measurable sections $\sigma_\iota(z)$, which form an orthonormal basis in $E(z)$ almost everywhere. Fixing such a choice, we can identify any measurable section of E by a column vector. Then $K_{M,E}(z, w)$ becomes a Hermitian matrix function on M. This is very convenient for studying the L^p-properties of the Bergman kernel. For instance, with this convention, the expression $K_{M,E}(z, w) f(w)$ in (7.2.5.1) becomes an ordinary matrix product.

However this trivialization is not very useful if the holomorphic aspects are to be considered.

7.4 LEMMA. *Notation and assumptions as in (7.1). Then the functions* $K(z, w) : w \in M$ *span a dense subspace of* $(L^2 \cap H)(M, E)$.

Proof. Otherwise there is a nonzero linear functional $\phi : (L^2 \cap H)(M, E) \to \mathbb{C}$, which kills all the sections $K(z, w) : w \in M$. ϕ corresponds to a dot product with a function $h \in (L^2 \cap H)(M, E)$. Thus $T_E h$ is identically zero, which is impossible since T_E is the identity on $(L^2 \cap H)(M, E)$. ∎

7.5 Definition-Proposition. The restriction of $K_{M,E}$ to the diagonal of $M \times M$ is called the corresponding *Bergman-type metric* on E:

$$K_{M,E}(z) = \sum u_\iota(z) \otimes \bar{u}_\iota(z).$$

$K_{M,E}(z)$ is a section of $\operatorname{Re}(E \otimes \bar{E})$. Thus it can be identified with a positive semidefinite Hermitian metric on E^*.

$K_{M,E}(z)$ is a positive definite Hermitian metric on E^* iff E is generated by its holomorphic L^2 sections.

Proof. Pick a linear map $w : E(z) \to \mathbb{C}$. Then

$$K_{M,E}(z)(w, \bar{w}) = \sum_{\iota} |w(u_\iota(z))|^2.$$

Thus $K_{M,E}(z)(w, \bar{w}) = 0$ iff the value of every holomorphic L^2 section, of E at z is contained in $\ker w$. ∎

7.5.1 Complement. If E is generated by its holomorphic L^2 sections, then $K_{M,E}(z)$ defines a duality between E and E^*; thus we obtain a Hermitian metric on E itself.

7.5.2 Special case. The Bergman kernel is especially interesting if E is the canonical bundle K_M. (7.1.1.3) gives a canonical choice for H.

$K_{M,K_M}(z)$ is a plurisubharmonic volume form on M that is canonically associated to M. In particular, it is invariant under biholomorphisms of M.

7.5.3 Note on terminology. If the canonical bundle admits a trivialization (e.g., when M is an open subset of $\mathbb{C}^n$), then $K_{M,K_M}(z)$ can be viewed as a Hermitian metric on $\mathcal{O}_M$.

The traditional notion of Bergman metric is the Ricci curvature of $K_{M,K_M}(z)$.

7.6 PROPOSITION. *Notation as above. Assume that M is connected. Then $K_{M,E}(z, z)$ and the Bergman kernel $K_{M,E}(z, w)$ determine each other.*

More generally, if $K'(z, w)$ is a section of $\pi_1^* E \otimes \pi_2^* \bar{E}$ that is holomorphic in z, antiholomorphic in w, and $K'(z, z) = K_{M,E}(z, z)$, then $K'(z, w) = K_{M,E}(z, w)$.

Proof. By definition, the kernel determines the metric. Since $K_{M,E}$ is holomorphic in z and antiholomorphic in w, it is sufficient to prove that $K_{M,E}(z, w)$ is determined by $K_{M,E}(z, z)$ in a neighborhood of the diagonal.

Pick a point $p \in M$ with local coordinates $z_1, \ldots, z_n$. Assume for notational simplicity that E is a line bundle, which we identify with the trivial bundle near p. On $M \times M$ we introduce local coordinates near (p, p) by the rules

$$x_\iota = (z_\iota, 0), \quad \text{and} \quad y_\iota = (0, \bar{z}_\iota).$$

The complex structure defined by these is not equivalent to the original one. The advantage is that $K_{M,E}$ becomes holomorphic in (x_ι, y_ι). The equations of the diagonal $\Delta \subset M \times M$ become $x_\iota = \bar{y}_\iota$. In the coordinate system

$$u_\iota = x_\iota + y_\iota, \quad v_\iota = \sqrt{-1}(x_\iota - y_\iota)$$

the equations of Δ become $\mathrm{im}(u_\iota) = \mathrm{im}(v_\iota) = 0$. A holomorphic function is determined by its restriction to the reals, which implies that $K_{M,E}(z, w)$ is determined by its restriction to the diagonal. ∎

7.7 Example. Assume that a group G of biholomorphisms acts transitively on M and that the action lifts to an action on E. Assume for simplicity that E is a line bundle and let h be a G-invariant metric on E. (h exists iff the representation of the stabilizer G_z of a point $z \in M$ in $GL(E(z))$ has compact image.) Assume also that E is generated by L^2 holomorphic sections. Then $K_{M,E}(z, z)$ is a nondegenerate G-invariant Hermitian metric on E^*. Therefore $K_{M,E}(z, z)$ is a constant multiple of the G-invariant metric h^*.

E^k is also generated by L^2 holomorphic sections (5.14), thus $K_{M,E^k}(z, z)$ is also a constant multiple of $(h^*)^k$. Applying (7.6) we obtain that there is a constant $c(k)$ such that

(7.7.1) $$K_{M,E^k}(z, w) = c(k)K_{M,E}(z, w)^k.$$

This is one of the very few examples when the Bergman kernel can be computed.

We are ready to investigate the surjectivity of the Poincaré map.

7.8 Notation. Let X be a compact complex manifold and E a vector bundle on X. Choose a volume form dv on X and a Hermitian metric h on E. Let $p : \tilde{X} \to X$ denote the universal cover. As usual, set $\tilde{E} = p^*E$, $\tilde{h} = p^*h$ and $d\tilde{v} = p^*dv$. Let $T : L^2(\tilde{X}, \tilde{E}) \to (L^2 \cap H)(\tilde{X}, \tilde{E})$ be the orthogonal projection and $K(z, w)$ the corresponding Bergman kernel.

In general, the Poincaré map $P : (L^1 \cap H)(\tilde{X}, \tilde{E}) \to H^0(X, E)$ is not surjective (cf. (8.9.3)). Our aim is to establish a result which says that under suitable assumptions on the Bergman kernel $K(z, w)$ the Poincaré map is surjective. Unfortunately, there are very few cases when these conditions are known to be satisfied. The most important class where they hold are bounded symmetric domains.

The conditions are analytic in nature and depend only on $\tilde{X}$ (and not on X). In order to emphasize this part, we replace $\tilde{X}$ by M, $\tilde{E}$ by F and $\tilde{h}(\ ,\ )d\tilde{v}$ by $h'(\ ,\ )d\mu$ in their formulation.

7.9 Condition 1. The Bergman projection $T = T_F$ extends to bounded linear maps

$$T^1 : L^1(M, F) \to L^1(M, F) \quad \text{and}$$
$$T^\infty : L^\infty(M, F) \to L^\infty(M, F).$$

7.10 Comments. Compactly supported continuous sections are weakly dense in $L^p(M, F)$ for $1 \leq p < \infty$ and weakly $*$-dense in $L^\infty(M, F)$, thus the above extensions are unique. By continuity these extensions automatically inherit several of the good properties of T (we set $p = 1, \infty$):

(7.10.1) T^p is idempotent: $T^p \circ T^p = T^p$.

(7.10.2) The image of T^p lies in $(L^p \cap H)(M, F)$. Indeed, a compactly supported continuous L^p function is also L^2, hence $T^p f = Tf$ is holomorphic. It is also L^p, hence $T^p f$ lies in $(L^p \cap H)(M, F)$. Therefore if $g \in L^p(M, F)$ is arbitrary then $T^p g$ lies in the closure of $(L^p \cap H)(M, F)$, which is itself.

(7.10.3) If $f \in L^1(M, F)$ and $g \in L^\infty(M, F)$ then

$$\int_M h'(T^1 f, g) d\mu = \int_M h'(f, T^\infty g) d\mu.$$

7.11 Condition 2. T^∞ is reproducing on holomorphic sections. I.e., if $g \in (L^\infty \cap H)(M, F)$ then $T^\infty g = g$.

The above conditions are extracted from [Earle69], which treated the case when M is a bounded symmetric domain. It generalized earlier results of [Bers65; Ahlfors64; Bell66].

7.12 THEOREM. [Earle69] *Notation as above. Assume that conditions 1–2 are satisfied by $(\tilde{X}, \tilde{E})$. Then the Poincaré map*

$$P : (L^1 \cap H)(\tilde{X}, \tilde{E}) \to H^0(X, E)$$

is surjective.

Proof. Choose an open fundamental domain $D \subset \tilde{X}$ such that its boundary has measure zero. Let χ_D be its characteristic function.

Let $\phi_1, \ldots, \phi_r$ be an orthonormal basis of $H^0(X, E)$. Set $\tilde{\phi}_i = p^* \phi_i$, these are in $(L^\infty \cap H)(\tilde{X}, \tilde{E})$. $\chi_D \tilde{\phi}_j$ is in $L^1(\tilde{X}, \tilde{E})$, thus

$$T^1(\chi_D \tilde{\phi}_j) \in (L^1 \cap H)(\tilde{X}, \tilde{E}).$$

We show that $P(T^1(\chi_D\tilde{\phi}_j)) = \phi_j$. By (7.10.3) and (7.11),

$$\int_{\tilde{X}} \tilde{h}(T^1(\chi_D\tilde{\phi}_j), \tilde{\phi}_k)d\tilde{v} = \int_{\tilde{X}} \tilde{h}(\chi_D\tilde{\phi}_j, T^\infty(\tilde{\phi}_k))d\tilde{v}$$

$$= \int_{\tilde{X}} \tilde{h}(\chi_D\tilde{\phi}_j, \tilde{\phi}_k)d\tilde{v} = \int_X h(\phi_j, \phi_k)dv.$$

If $f \in (L^1 \cap H)(\tilde{X}, \tilde{E})$ and $\psi \in H^0(X, E)$ then

$$\int_X h(P(f), \psi)dv = \int_{\tilde{X}} \tilde{h}(f, p^*\psi)d\tilde{v}.$$

Combining these two equalities we obtain that

$$\int_X h(P(T^1(\chi_D\tilde{\phi}_j)), \phi_k)dv = \int_{\tilde{X}} \tilde{h}(T^1(\chi_D\tilde{\phi}_j), \tilde{\phi}_k)d\tilde{v}$$

$$= \int_X h(\phi_j, \phi_k)dv, \quad \forall j, k.$$

Therefore $P(T^1(\chi_D\tilde{\phi}_j)) = \phi_j$. ∎

It remains to find certain cases when conditions 1–2 are satisfied. The answer provided by (7.13) is very reasonable for condition 1. The assumptions of (7.14) are, however, quite artificial. It is not clear to me how restrictive condition 2 is in reality.

7.13 Proposition. *Notation as above. Assume that for every $w \in M$ the restriction $K(z, w) \in \Gamma(M, F)$ is L^1 and its norm is bounded by a constant (independent of w). Then the condition 1 is satisfied.*

Proof. Assume for simplicity that F is trivialized as in (7.3). The functional T becomes

$$Tf(z) = \int_M K(z, w)f(w)d\mu(w),$$

where $K(z, w)$ is a matrix function and $f(w)$ is a column vector. Assume now that f is in L^1. We show that $\|T^1 f\|_1 \le C\|f\|_1$, where C is the sum of the L^1 norms of the entries of $K(z, w)$.

In order to see this, it is sufficient to show that if h is an L^∞ row vector then

$$\int_M h(z)Tf(z)d\mu(z) = \int_{M\times M} h(z)K(z, w)f(w)d\mu(w)d\mu(z)$$

$$\le C\|f\|_1\|h\|_\infty.$$

This is, however, obvious by integrating first by z. The statement about T^∞ is clear. ∎

The method of [Earle69] requires more complicated assumptions about the Bergman kernel in order to get condition 2. To get something reasonable, assume that F is a line bundle. The Bergman kernel associated to F is denoted by $K_1(z, w)$ and the Bergman kernel associated to F^2 by $K_2(z, w)$.

7.14 PROPOSITION. *Notation as above. Assume that*

(7.14.1) there is a constant C such that $\|K_i(z, w_1)\|_1 < C$ for every $w_1 \in M$ and $i = 1, 2$;

(7.14.2) for every $w_1, w_2 \in M$ the section of M given by $K_2(z, w_1)/K_1(z, w_2)$ is L^2. (In particular $K_1(z, w)$ is nowhere zero.)

(7.14.3) $(L^1 \cap H)(M, F^2) \subset (L^2 \cap H)(M, F^2)$. (By (5.14) this is satisfied if Aut(M) *has a compact fundamental domain.)*

Then condition 2 is satisfied.

Proof. Fix $w \in M$ and let $L^1_w(M, F)$ be the space of measurable sections $f(z)$ such that $K_1(z, w)f(z)$ is an L^1 section of F^2. The definition of $L^1_w(M, F)$ is made so that multiplication by $K_1(z, w)$ provides an isometry

$$(7.14.4) \quad K_1(z, w) : L^1_w(M, F) \hookrightarrow L^1(M, F^2).$$

$K_1(w, z)$ is the conjugate of $K_1(z, w)$, thus the linear map T^1_w : $L^1_w(M, F) \to E(w)$ given by

$$T^1_w(f) = \int_M K_1(w, t)f(t)dv(t)$$

has norm 1. We prove that if f is holomorphic then

$$(7.14.5) \quad T^1_w(f) = f(w).$$

By (7.14.1) $L^\infty(M, F) \subset L^1_w(M, F)$, so this implies condition 2 if we vary w.

Evaluation at w is a continuous linear functional, thus it is sufficient to check (7.14.5) over a weakly dense subset of functions.

Set $f_u = K_2(z, u)/K_1(z, w)$ for $u \in M$. $f_u \in L^1_w(M, F)$ by assumption (7.14.1). By (7.14.2) f_u is L^2, hence $T^1_w(f_u) = (Tf_u)(w) = f_u(w)$ since T is the identity on L^2 holomorphic functions.

It remains to show that sections of the form $K_2(z, u)/K_1(z, w)$ span a weakly dense subspace of $(L^1_w \cap H)(M, F)$. Applying the isometry (7.14.4) we are reduced to show that $K_2(z,u) : \ u \in M$ span a weakly dense subspace of $(L^1 \cap H)(M, F^2)$.

Let $g \in L^\infty(M, F^2)$ be such that

$$\int h(K_2(z,u), g) d\mu(z) = 0 \quad \forall\, u \in M.$$

Then $T^\infty g = 0$, hence by (7.10.3) we conclude that

$$\int h(T^1 h, g) d\mu(z) = 0 \quad \forall\, h \in L^1(M, F^2).$$

Since $(L^1 \cap H)(M, F^2) \subset (L^2 \cap H)(M, F^2)$, T^1 is the identity on $(L^1 \cap H)(M, F^2)$. Thus any linear functional that is zero on all the $K_2(z,u)$ is zero on $(L^1 \cap H)(M, F^2)$. ∎

7.15 PROPOSITION. *Assume that M is homogeneous and F is a line bundle. Let G be a group of biholomorphisms acting transitively on M. Assume that the action lifts to an action on F. Let h be a G-invariant metric on F. Assume that F is generated by L^2 holomorphic sections.*

(7.15.1) Then condition 1 is satisfied for (M, F^k) for $k \geq 2$.

(7.15.2) If for every $w_1, w_2 \in M$ the function $K_{M,F^k}(z, w_1)/K_{M,F^k}(z, w_2)$ is bounded on M, then condition 2 is staisfied for (M, F^k) $(k \geq 2)$.

Proof. By (7.7), in this case $K_{M,F^k}(z,w) = c(k) K_{M,F}(z,w)^k$ for some constant $c(k)$. By (7.2.1), $K_{M,F}(z,w)|M \times \{w\}$ is L^2 for fixed w, thus $K_{M,F}(z,w)^k|M \times \{w\}$ is L^1 for $k \geq 2$. By homogeneity, the L^2 norm of $K_{M,F}(z,w)|M \times \{w\}$ is independent of w. This shows the first part.

By (7.7.1)

$$K_{M,F^{2k}}(z, w_1)/K_{M,F^k}(z, w_2) = \frac{c(2k)}{c(k)} K_{M,F^k}(z, w_1)/K_{M,F^k}(z, w_2),$$

thus (7.15.2) implies (7.14.2). ∎

7.15.3 Remark. The assumptions of (7.15) are easy to check if the Bergman kernel is known explicitly, e.g., for the classical bounded symmetric domains. For the exceptional cases it was verified in [Resnikoff69]. The case of the ball is worked out in (8.9.4.1).

7.16 Comment. Under the assumptions of (7.15) the spaces $L^1_{w_1}(M, F)$ and $L^1_{w_2}(M, F)$ are identical (and isomorphic as Banach spaces). Thus $L^1_w(M, F)$ is the natural existence domain of the projection T. T is then reproducing on holomorphic functions whenever it is defined.

It is not much extra work to prove the surjectivity criterion for Poincaré maps in the noncompact quotient case as well.

7.17 Notation. Let M be a complex space with a holomorphic line bundle F. Let Γ be a discontinuous group of biholomorphisms of (M, F). (We do not assume that the action is free.) Fix an invariant measure $d\mu$ and an invariant Hermitian metric h on F. Let $D \subset M$ be an open fundamental domain for Γ such that its boundary has measure zero.

If f is a Γ-invariant measurable section of F then we can define its $(L^p)^\Gamma$-norm by

$$\|f\|_{p,\Gamma} = \left(\int_D h(f, \bar{f})^{p/2} d\mu\right)^{1/p}.$$

We use the notation $(L^p)^\Gamma(M, F)$ (resp. $(L^p \cap H)^\Gamma(M, F)$ to denote the space of Γ-invariant sections of F whose $(L^p)^\Gamma$-norm is finite (resp. and are also holomorphic).

Let $K_s(z, w)$ denote the Bergman kernel associated to (M, F^s).

7.18 THEOREM. [Earle69] *Notation as above. Assume that*

(7.18.1) for every $w \in M$ the restriction $K(z, w) \in \Gamma(M, F)$ is L^1 and its norm is bounded by a constant C (independent of w);

(7.18.2) for every $w_1, w_2 \in M$ the section of F given by $K_2(z, w_1)/K_1(z, w_2)$ is L^2.

Then the Poincaré map

$$P : (L^1 \cap H)(M, F) \to (L^1 \cap H)^\Gamma(M, F)$$

is surjective.

Proof. The proof is essentially the same as for (7.12). Pick any $\phi \in (L^1 \cap H)^\Gamma(M, F)$. We would like to show that

$$P(T^1(\chi_D \phi)) = \phi.$$

Let $\psi \in (L^\infty \cap H)^\Gamma(M, F)$. As in the proof of (7.12) we intend to use the chain of equalities:

$$\begin{aligned} \int_D h(P(T^1(\chi_D\phi)), \psi) d\mu &= \int_M h(T^1(\chi_D\phi), \psi) d\mu \\ &= \int_M h(\chi_D\phi, T^\infty\psi) d\mu \qquad (7.18.3) \\ &= \int_D h(\phi, \psi) d\mu. \end{aligned}$$

These equalities use only conditions 1–2, hence they hold by (7.13–14).

The slight complication is that (7.18.3) implies (7.18) only if $(L^\infty \cap H)^\Gamma(M, F)$ is dense in the dual space of $(L^1 \cap H)^\Gamma(M, F)$. This requires a further argument.

Let $S : (L^1 \cap H)^\Gamma(M, F) \to \mathbb{C}$ be a bounded linear functional. We can extend it to $\bar{S} : (L^1)^\Gamma(M, F) \to \mathbb{C}$, thus it is represented by integration against a section $g \in (L^\infty)^\Gamma(M, F)$. Our hope is that $T^\infty g$ represents S. Let $f \in (L^1 \cap H)^\Gamma(M, F)$. We need to justify the equalities

$$\text{(7.18.4)} \quad \int_D h(T^\infty g, f) d\mu = \int_D h(g, Tf) d\mu = \int_D h(g, f) d\mu.$$

In general f is not in $L^1(M, F)$, thus it is not clear that T can be extended to $(L^1)^\Gamma(M, F)$ and that it is reproducing on holomorphic functions there. However both of these follow if we prove that

$$\text{(7.18.5)} \quad (L^1)^\Gamma(M, F) \subset L^1_w(M, F) \quad \text{for almost every } w \in M,$$

where $L^1_w(M, F)$ was defined in the proof of (7.14). By the Γ-invariance, it is sufficient to prove (7.18.5) for almost every $w \in D$. This follows from the inequality:

$$\begin{aligned}
&\int_D d\mu(z) \int_M |K(z, w)||f(w)| d\mu(w) \\
&= \int_D d\mu(z) \sum_\Gamma \int_D |K(z, \gamma w)||f(w)| d\mu(w) \\
&= \int_D d\mu(z) \sum_\Gamma \int_D |K(\gamma^{-1} z, w)||f(w)| d\mu(w) \\
&= \int_D |f(w)| d\mu(w) \int_M |K(z, w)| d\mu(z) \\
&\le C \int_D |f(w)| d\mu(w).
\end{aligned}$$

Thus all the steps in (7.18.4) are justified. ∎

CHAPTER 8

Ball Quotients

In order to illustrate some of the previous methods, we compute in detail the example of the unit ball in $\mathbb{C}^n$. This is the easiest example of a Hermitian homogeneous space.

8.1 Definition. Consider the standard action of $SL(n+1,\mathbb{C})$ on $\mathbb{CP}^n$. We write coordinates on $\mathbb{CP}^n$ as $(x_0 : \ldots : x_n)$. On the affine chart $x_0 \neq 0$ we introduce affine coordinates $z_i = x_i/x_0$ $(i \neq 0)$. Let $SU(1,n) < SL(n+1,\mathbb{C})$ be the subgroup that leaves the Hermitian form $Q = -x_0\bar{x}_0 + x_1\bar{x}_1 + \cdots + x_n\bar{x}_n$ invariant. In the above affine chart $Q = 0$ can be written as

$$|z_1|^2 + \cdots + |z_n|^2 = 1,$$

which is the equation of the unit sphere in $\mathbb{C}^n$. It is not too hard to see that the action of $SU(1,n)$ on $\mathbb{CP}^n$ has three orbits: the unit sphere $S(1)$, the unit ball $B = B(1)$, and their complement, where we use the notation

$$B(r) := \{(z_1, \ldots, z_n) \in \mathbb{C}^n |\ |z_1|^2 + \cdots + |z_n|^2 < r^2\}, \quad \text{and}$$
$$S(r) := \{(z_1, \ldots, z_n) \in \mathbb{C}^n |\ |z_1|^2 + \cdots + |z_n|^2 = r^2\}.$$

Let $\mathcal{O}(1)$ be the tautological line bundle on $\mathbb{CP}^n$: the homogeneous coordinates x_i are sections of $\mathcal{O}(1)$. By definition, there is a natural $SL(n+1,\mathbb{C})$-action on $\mathcal{O}(1)$. Let M^{-1} be the restriction of $\mathcal{O}(1)$ to B. The reason for this choice of M becomes clear later. In order to avoid confusion in notation, we denote by $\sigma^{-1}(z)$ or by σ^{-1} the restriction of the section x_0 of $\mathcal{O}(1)$ to a section of M^{-1}. σ^{-1} is a nowhere zero holomorphic section which one would usually think of as the constant section 1. In particular, M^{-1} is isomorphic to the trivial line bundle on B.

M^{-1} comes equipped with an $SU(1,n)$-action. We can also take the tensor powers M^k : $k \in \mathbb{Z}$ with their induced $SU(1,n)$-action. Taking k^{th}-tensor power gives a nowhere zero section σ^k of M^k. The M^k are all isomorphic as line bundles on B, but as we will see in a moment, they are different as line bundles with $SU(1,n)$-action.

8.2 CLAIM. *There is a unique (up to a multiplicative constant) $SU(1,n)$-invariant Hermitian metric on M. Suitably normalized it is*

$$\|\sigma(z)\| = (1-|z|^2)^{1/2}.$$

Proof. A metric h on M^{-1} can be identified with an everywhere real section h' of $M^{-1}\otimes(\overline{M})^{-1}$, where $(\overline{M})^{-1}$ is the conjugate antiholomorphic line bundle. h is invariant under a group action iff h' is invariant under the corresponding group action. By our definition

$$(8.2.1)\quad -\sigma^{-1}\otimes\bar{\sigma}^{-1}\cdot(1-|z|^2) = -x_0\otimes\bar{x}_0 + x_1\otimes\bar{x}_1 + \cdots x_n\otimes\bar{x}_n$$

is $SU(1,n)$-invariant. Thus $\|\sigma^{-1}(1-|z|^2)^{1/2}\|$ is constant in z. We can assume that it is 1. The sign is changed to + to get a positive definite metric. ∎

8.3 LEMMA. *Notation as above. Then*

$$K_B \cong M^{n+1} \quad \text{as } SU(1,n)\text{-line bundles.}$$

Therefore

$$(\sqrt{-1})^n(1-|z|^2)^{-(n+1)}dz_1\wedge d\bar{z}_1\wedge\cdots\wedge dz_n\wedge d\bar{z}_n$$

is the unique (up to a multiplicative constant) $SU(1,n)$-invariant volume form on B.

Proof. We know that $K_{\mathbb{CP}^n}\cong\mathcal{O}(-n-1)$ as line bundles with $SL(n+1,\mathbb{C})$ action. Under this isomorphism $dz_1\wedge\cdots\wedge dz_n$ corresponds to $x_0^{-(n+1)}$. By restriction we obtain the first claim. Combining this with (8.2) we obtain that the invariant volume form is

$$\text{const}\cdot(1-|z|^2)^{-(n+1)}\sigma^{n+1}\wedge\bar{\sigma}^{n+1},$$

which gives the second part. ∎

We are now ready to investigate the existence of L^p-sections of M^k. (8.2–3) allow us to translate the L^p condition to a more conventional integral:

8.4 PROPOSITION. *Let f be a holomorphic function on B and $f\cdot\sigma^k$ the corresponding section of M^k. Then $f\cdot\sigma^k$ is L^p (with respect to the above invariant metric $\|\ \|$) iff*

$$\int_B |f|^p(1-|z|^2)^{(pk/2)-n-1}dm < \infty,$$

where dm is the standard Lebesgue measure. ∎

In order to see what this means, we need an easy lemma.

8.5 LEMMA. *Let $\phi(r)$ be a continuous, positive real function on $[0, 1)$ and g a nonzero plurisubharmonic function on B. Then*

$$\int_B \phi(|z|)g(|z|)dm < \infty \quad \Rightarrow \quad \int_B \phi(|z|)dm < \infty.$$

Proof.

$$\int_B \phi(|z|)g(|z|)dm = \int_0^1 \phi(r) \int_{S(r)} |g(z)|.$$

$\int_{S(r)} |g(z)|$ is an increasing function of r since g is plurisubharmonic. ∎

8.6 COROLLARY. *Notation as in (8.4).*

(8.6.1) Assume $p > 0$ and $pk/2 - n - 1 \leq -1$. Then every L^p-section of M^k is identically zero.

(8.6.2) If f is bounded on B then $f \cdot \sigma^k$ is L^p iff $pk/2 - n - 1 > -1$.

Proof. Apply (8.5) with $g = |f|^p$ and $\phi = (1 - r^2)^{pk/2-n-1}$. Easy computation shows that $\int_D \phi = \infty$ iff $pk/2 - n - 1 \leq -1$. By (8.5) this proves (8.6.1) and (8.6.2) follows directly. ∎

8.7 COROLLARY. *Notation as above.*

(8.7.1) M^k has nonzero L^2-sections iff $k \geq n + 1$;

(8.7.2) M^k has nonzero L^1-sections iff $k \geq 2(n + 1) - 1$. ∎

We want to apply the above results to study varieties whose universal cover is B. It is not at all clear that $B \subset \mathbb{C}^n$ occurs as a universal cover of algebraic varieties for $n \geq 2$. The first examples are given in [Borel63]. Below is a short summary of the various methods that have been developed to find such varieties.

8.8 Examples of ball quotients. The stabilizer of any point in $SU(1, n)$ is compact. Thus if $\Gamma < SU(1, n)$ is a subgroup, then Γ acts on B properly and with a compact fundamental domain iff Γ is discrete in $SU(1, n)$ and $\Gamma \backslash SU(1, n)$ is compact. I know of four ways of constructing such subgroups.

(8.8.1) $SU(1, 1)$. This is rather special. By the uniformization theorem, if C is a compact Riemann surface of genus at least 2, then the universal cover of C is the unit disc. Thus we obtain plenty of examples.

(8.8.2) Arithmetic subgroups. The natural generality of this method is for arbitrary semisimple real Lie groups [Borel63]. The statement of the general result would take us too far. Instead, let me just indicate the construction of certain subgroups $\Gamma < SU(1,n)$ without proving that they have the required properties.

Pick a positive square-free integer d and let $\mathbb{Z}[\sqrt{d}]$ be the set of real numbers of the form $a+b\sqrt{d}$, where $a, b \in \mathbb{Z}$. (For theoretical purposes it would be better to look at all algebraic integers of the field $\mathbb{Q}(\sqrt{d})$.) Let Q_d be the Hermitian form

$$Q_d = -\sqrt{d}x_0\bar{x}_0 + x_1\bar{x}_1 + \cdots + x_n\bar{x}_n.$$

Let $SU(Q_d) < SL(n+1, \mathbb{C})$ be the subgroup that leaves the Hermitian form Q_d invariant. $SU(1,n)$ and $SU(Q_d)$ are conjugate inside $GL(n+1, \mathbb{C})$, thus $SU(1,n)$ and $SU(Q_d)$ are isomorphic as real Lie groups.

Let $SU(Q_d, \mathbb{Z}[\sqrt{d}]) < SU(Q_d)$ be the subgroup of matrices with entries in $\mathbb{Z}[\sqrt{d}]$. One can see that $SU(Q_d, \mathbb{Z}[\sqrt{d}]) < SU(Q_d)$ is discrete and cocompact. Thus $SU(Q_d, \mathbb{Z}[\sqrt{d}])$ acts on B properly and with compact fundamental domain. Also, a suitable subgroup of finite index $\Gamma < SU(Q_d, \mathbb{Z}[\sqrt{d}])$ acts freely. Thus by (5.22) $\Gamma\backslash B$ is a projective algebraic variety.

(8.8.3) [Yau77] proves that a smooth projective variety X^n is a quotient of the ball $B \subset \mathbb{C}^n$ iff K_X is ample and

$$nc_1(X)^n = 2(n+1)c_1(X)^{n-2}c_2(X).$$

In particular, for $n = 2$ this becomes $c_1^2 = 3c_2$. If one can construct a smooth projective variety X satisfying the above conditions, then X is a ball quotient, though one may know very little about the fundamental group itself. This approach has been successful in dimension 2. [Mumford79] constructed a ball quotient X such that $b_2(X) = 1$ and $h^0(X, K_X) = 0$. It is interesting to note that his construction uses algebraic varieties over a p-adic field. His example was further clarified by [Ishida88].

Further examples were constructed subsequently. The simplest ones can be found in [Hirzebruch83, 3.2].

(8.8.4) A few examples are constructed in [Deligne-Mostow86] for small values of n.

8.9 Automorphic forms on ball quotients. Let X be a smooth projective variety such that its universal cover is biholomorphic to the ball $B \subset \mathbb{C}^n$. Set $\Gamma = \pi_1(X)$. Γ acts by deck transformations on B; thus we obtain a natural embedding $\Gamma < \mathrm{Aut}(B)$. It is important to note that $\mathrm{Aut}(B) = PSU(1,n)$ (see, e.g., [Rudin80, 2.1.3]), which is the quotient of $SU(1,n)$ by the diagonal matrices $\epsilon^j I$, where ϵ is an $(n+1)^{st}$ root of unity and I is the identity matrix. In general we cannot lift the embedding $r : \Gamma \to PSU(1,n)$ to an embedding $\bar{r} : \Gamma \to SU(1,n)$.

Assume, however, that it is possible to find a lifting $\bar{r}$. (One can see that this is always possible by passing to a suitable finite index subgroup of Γ.) Let $\rho : \Gamma \to \mathbb{Z}_{n+1}$ be a character. Then

$$(\rho\bar{r})(\gamma) = \epsilon^{\rho(\gamma)}\bar{r}(\gamma)$$

defines another lifting. One can easily see that the liftings form a principal homogeneous space under $\operatorname{Hom}(\Gamma, \mathbb{Z}_{n+1})$.

Once a lifting $\bar{r}$ is fixed, we obtain a lifting of the action of Γ on B to an action on M. Therefore M descends to a line bundle M_X on X such that $M_X^{n+1} \cong K_X$. The latter condition specifies M_X up to an $(n+1)$-torsion element of $\operatorname{Pic}(X)$. These are again in one-to-one correspondence with $\operatorname{Hom}(\Gamma, \mathbb{Z}_{n+1})$.

We want to study the vector spaces of automorphic forms

$$H^0(X, M_X^m) \quad \text{for } m \geq 1.$$

Four cases should be distinguished.

(8.9.1) $1 \leq m \leq n$. In this range the general methods say nothing. If $n = 1$ then $X = C$ is a curve and M_C is called a "theta characteristic." Even for fixed C, the dimension of $H^0(C, M_C)$ depends on the choice of M_C.

(8.9.2) $m = n+1$. $M_X^{n+1} = K_X$, thus we would like to know $H^0(X, K_X)$. One expects that usually this is positive dimensional, in fact quite large. The example of [Mumford79] mentioned in (8.8.3) shows that it can be zero. At least for surfaces, $H^0(X, K_X) = 0$ iff $\chi(X, K_X) = 1$ and there are only finitely many such surfaces up to isomorphism. The same might be true in all dimensions.

(8.9.3) $n+2 \leq m \leq 2n$. Set $m = n+1+r$ and write $M_X^m = K_X \otimes M_X^r$. By (6.4) and (8.7.1) we obtain that

$$H^0(X, K_X \otimes M_X^r) > 0 \quad \text{for } r \geq 1.$$

Thus we have nonzero automorphic forms of weight m. On the other hand, by (8.7.2) M^m has no nonzero L^1 sections. Thus none of the automorphic forms of weight m can be represented by Poincaré series.

(8.9.4) $m \geq 2n+1$. As before, we see that there are nonzero automorphic forms. Also, M^m has nonzero L^1 sections. In contrast with the previous case, the Poincaré series map is surjective.

8.9.4.1 THEOREM. [Earle69] *Notation as above. Let $m \geq 2n+1$. The Poincaré series map*

$$P : (L^1 \cap H)(B, M^m) \to H^0(X, M_X^m) \quad \textit{is surjective.}$$

Proof. We need to check that the Bergman kernel $K_{B,m}(z, w)$ of M^m satisfies the conditions (7.13–14). The ball is homogeneous, thus the L^1-norm of $K_{B,m}(z, w)|B \times \{w\}$ is independent of w. The computation of the Bergman kernel (8.10.2) shows that there is a universal constant $c(m)$ such that

$$(1+|w|)^{-m} \leq c(m)\frac{K_{B,m}(z, w)}{\sigma(z)^m \otimes \bar{\sigma}(w)^m} \leq (1-|w|)^{-m},$$

where for simplicity we write $\sigma(z)$ instead of $\pi_1^*\sigma$ and $\sigma(w)$ instead of $\pi_2^*\sigma$. By (8.2) and (8.6) we obtain that $K_{B,m}(z, w)|B \times \{w\}$ is L^1 if $m \geq 2n+1$. Also the quotients $K_{B,2m}(z, w_1)/K_{B,m}(z, w_2)$ are L^2 (even L^1). ∎

8.10 Computation of the Bergman kernel of the ball. The ball is homogeneous and the $SU(1, n)$ action lifts to M^m. Thus by (7.7) the restriction of the Bergman kernel to the diagonal is a multiple of the invariant Hermitian metric

$$(8.10.1) \quad K_{B,m}(z) = c(m)(1-|z|^2)^{-m}\sigma(z)^m \otimes \bar{\sigma}(z)^m.$$

Using (7.6) and a good guess we obtain the Bergman kernel

$$(8.10.2) \quad K_{B,m}(z, w) = c(m)(1-\langle z, w\rangle)^{-m}\sigma(z)^m \otimes \bar{\sigma}(w)^m,$$

where $\langle z, w\rangle = \sum z_i\bar{w}_i$.

We can write a section of M^m as $f\sigma^m$. The corresponding Bergman projection T_m is given by

$$T_m(f)(z)\sigma(z)^m = c(m)\int_B \sigma(z)^m \otimes \bar{\sigma}(w)^m(1-\langle z, w\rangle)^{-m} f\sigma(w)^m d\mu(w).$$

Taking into account that $\|\bar{\sigma}(w)^m \cdot \sigma(w)^m\| = (1-|w|^2)^m$ we obtain that

$$(8.10.3) \quad T_m(f)(z) = c(m)\int_B \left(\frac{1-|w|^2}{1-\langle z, w\rangle}\right)^m f(w)\frac{dm}{(1-|w|^2)^{n+1}},$$

where dm is the standard Lebesgue measure. These are the Bergman-type projections as considered in [Rudin80, 7.1]. When M^m is the canonical bundle, i.e., $m = n + 1$, we obtain the shorter expression

$$(8.10.4) \quad T_{n+1}(f)(z) = c(n+1) \int_B (1 - \langle z, w \rangle)^{-(n+1)} f(w) dm(w).$$

This is the traditional Bergman kernel on the ball (see, e.g., [Rudin80, 3.1]). It is not hard to compute the constant [Rudin80, 7.1.1.(4)]

$$(8.10.5) \quad c(m) = \frac{1}{\operatorname{vol} B_n} \binom{n+m}{n} = \frac{n!}{\pi^n} \binom{n+m}{n}.$$

It is interesting to look at the example of some noncompact quotients as well.

8.11 Example. Let $\bar{C}$ be a smooth compact Riemann surface and $B \subset \bar{C}$ a finite set of points. Let $C = \bar{C} \setminus B$. Assume that the universal cover of C is the unit disc Δ. This means that $|B| \geq 3$ if $\bar{C} \cong \mathbb{CP}^1$ and $|B| \geq 1$ if $\bar{C}$ is elliptic.

One example that has been studied classically in connection with the little Picard theorem is the case $\bar{C} \cong \mathbb{CP}^1$ and $B = \{0, 1, \infty\}$ (see, e.g., [Rudin66, 16.17–22]). It is possible to give a very nice explicit description of the universal cover. We replace Δ with the upper half plane $\mathbb{H}$. Let

$$\sigma(z) = \frac{z}{2z+1} \quad \text{and} \quad \tau(z) = z + 2$$

be two elements of $PSL(2, \mathbb{Z})$ acting on $\mathbb{H}$. One can see that they generate a free group, which is the subgroup of $PSL(2, \mathbb{Z})$ given as

$$\Gamma_2 = \left\{ \begin{pmatrix} a & b \\ c & d \end{pmatrix} \in PSL(2, \mathbb{Z}) \;\middle|\; \begin{pmatrix} a & b \\ c & d \end{pmatrix} \equiv \begin{pmatrix} 1 & 0 \\ 0 & 1 \end{pmatrix} \bmod 2 \right\}.$$

It is easy to see that $\Gamma_2 \backslash \mathbb{H} \cong \mathbb{CP}^1 \setminus \{0, 1, \infty\}$.

In all cases, the invariant metric on Δ descends to a metric $\| \;\|$ on C with Gaussian curvature -1. If z is a local coordinate at a point $P \in B$, then near P we have

$$c^{-1}|z|(-2\log|z|) \leq \|dz\| \leq c|z|(-2\log|z|),$$

for some $c \geq 1$. Let D be a small disc around P. $z^m(dz)^n$ is an L^p-section of ω_C^n locally near P iff

$$\int_D |z|^{pm}|z|^{pn}(-2\log|z|)^{pn} \frac{dz \wedge d\bar{z}}{|z|^2(-2\log|z|)^2} < \infty.$$

Assuming that $pn \geq 1$ this is equivalent to $m \geq 1 - n$. Thus we obtain the following.

8.11.1 CLAIM. *Notation as above. Let $p \geq 1$. Then*

$$(L^p \cap H)(C, \omega_C^n) = H^0(\bar{C}, \omega_{\bar{C}}^n((n-1)B)). \quad \blacksquare$$

In contrast with the case of compact quotients we obtain the following.

8.11.2 COROLLARY. *Let $\bar{C} \cong \mathbb{CP}^1$ and $B = \{0, 1, \infty\}$. Then ω_C^2 has no L^1 or L^2 sections.* $\blacksquare$

This also gives an example when all the Poincaré series are zero, though the space of L^1-sections of ω_Δ^2 is infinite dimensional.

8.12 Hirzebruch proportionality principle. This is a method that can be used to compute $h^0(X, K_X \otimes M_X^r)$ for ball quotients, and more generally for quotients of bounded symmetric domains.

8.12.1 THEOREM. [Hirzebruch58] *Let $X = \Gamma \backslash B$ be a ball quotient as above such that M_X exists. Then*

$$h^0(X, K_X \otimes M_X^r) = \chi(X, K_X)\binom{r+n}{n} \quad \textit{for } r \geq 1,$$

and $\chi(X, K_X)$ is positive.

Proof. Let me just outline the main steps of the proof. Our aim is to compare

$$\chi(\Gamma \backslash B, M_X^r) \quad \text{and} \quad \chi(\mathbb{CP}^n, \mathcal{O}(-1)^r).$$

First we observe that $c_1(M_X) = c_1(X)/(n+1)$ and $c_1(\mathcal{O}(-1)) = c_1(\mathbb{CP}^n)/(n+1)$. Thus in the Hirzebruch-Riemann-Roch theorem (6.1.1) we can express the Chern forms of the line bundle in terms of the Chern forms of the underlying variety. Expanding by the powers of r we obtain that

$$\begin{aligned} &\chi(X, M^r) = \sum_i r^i \int T_i(c_1(X), \ldots, c_n(X)), \quad \text{and} \\ (8.12.2)\qquad &\chi(\mathbb{CP}^n, \mathcal{O}(-r)) = \sum_i r^i \int T_i(c_1(\mathbb{CP}^n), \ldots, c_n(\mathbb{CP}^n)), \end{aligned}$$

where the polynomials $T_i(x_1, \ldots, x_n)$ depend only on n and $c_i(Z)$ is the ith Chern form of Z (in some metric).

B is homogeneous under the group $SU(1, n)$ and $\mathbb{CP}^n$ is homogeneous under the group $SU(n+1)$. In both cases we can choose an invariant metric on the tangent bundle and compute the Chern forms using this metric. The Chern form c_i is then also invariant under the group action. Therefore it is sufficient to compute c_i at one point. Let our distinguished point be $P = (1 : 0 : \dots : 0)$ with local affine coordinates $z_1, \dots, z_n$. Let $V = \langle dz_1, \dots, dz_n \rangle$ be the cotangent space at P. The curvature tensors $R(B^n)$ of B^n at P and $R(\mathbb{CP}^n)$ of $\mathbb{CP}^n$ at P are elements of the same vectorspace $V \otimes \bar{V}^* \otimes V \otimes \bar{V}$.

The computation of the curvature tensors can be done in terms of the Lie algebra of the corresponding groups [Helgason78, IV.4.2], and we obtain that

$$(8.12.3) \quad R(B^n) = -R(\mathbb{CP}^n).$$

Let $d\mu(Z)$ denote the invariant volume form ($Z = B^n$ or $Z = \mathbb{CP}^n$) which corresponds to a fixed volume form on V. $T_i(c_1(Z), \dots, c_n(Z))$ is also an invariant (n, n) form, thus there are constants $t_i(Z)$ such that

$$T_i(c_1(Z), \dots, c_n(Z)) = t_i(Z) d\mu(Z).$$

We can descend $c_i(B)$ and $d\mu(B)$ to X, and therefore

$$T_i(c_1(X), \dots, c_n(X)) = t_i(B) d\mu(X).$$

Therefore $t_i(X) = t_i(B)$ and using (8.12.3) we see that

$$(8.12.4) \quad t_i(X) = (-1)^n t_i(\mathbb{CP}^n).$$

Substituting into (8.12.2) we obtain the equalities

$$(8.12.5) \quad \begin{aligned} \frac{1}{\operatorname{vol} \mathbb{CP}^n} \chi(\mathbb{CP}^n, \mathcal{O}(r)) &= \sum (-r)^i p_i(\mathbb{CP}^n) \\ &= (-1)^n \sum (-r)^i p_i(B^n) \\ &= (-1)^n \frac{1}{\operatorname{vol} X} \chi(X, M_X^{-r}) \\ &= \frac{1}{\operatorname{vol} X} \chi(X, K_X \otimes M_X^r). \end{aligned}$$

It is easy to compute that

$$\chi(\mathbb{CP}^n, \mathcal{O}(r)) = \binom{n+r}{n}.$$

Setting $r = 0$ we obtain that $\chi(X, K_X) = \operatorname{vol} X/\operatorname{vol} \mathbb{CP}^n$. Therefore (8.12.5) becomes

$$\chi(X, K_X \otimes M_X^r) = \chi(X, K_X)\binom{r+n}{n}.$$

Combining this with the Kodaira vanishing theorem (9.1) gives (8.12.1). The positivity of $\chi(X, K_X)$ follows, since for $r \gg 1$ the left-hand side of the formula is positive. ∎

Part III

Vanishing Theorems

CHAPTER 9

The Kodaira Vanishing Theorem

In this chapter I will present a simple proof of the Kodaira vanishing theorem and some of its generalizations. More subtle vanishing theorems will be treated in subsequent chapters.

9.1 THEOREM. [Kodaira53]

(9.1.1) Let M be a compact complex manifold and L a line bundle on M with a Hermitian metric h whose curvature form is positive definite everywhere. Then

$$H^i(M, K_M \otimes L) = 0 \quad \text{for } i > 0.$$

(9.1.2) Let X be a smooth projective variety and L an ample line bundle on X. Then

$$H^i(X, K_X \otimes L) = 0 \quad \text{for } i > 0.$$

The two versions are equivalent, but this misses the point. It is fairly easy to see that an ample line bundle carries a Hermitian metric h whose curvature form is positive definite everywhere [Griffiths-Harris78, p. 148]. Thus (9.1.1) $\Rightarrow$ (9.1.2). Conversely, the assumptions in (9.1.1) imply that M is projective algebraic and L is ample. Kodaira, however, used (9.1.1) in order to prove this. We concentrate on the second form of the theorem, which is unfortunately weaker.

9.2 Convention. For the rest of this section, all cohomologies are in the Euclidean topology.

The proof follows the general philosophy outlined in [Kollár86b, p. 5]. According to that approach, a vanishing theorem follows if there is a topological sheaf G, a coherent sheaf F, and a morphism $G \to F$ that is surjective on cohomologies.

The surjectivity result that I need is the following basic result of Hodge theory. For the proof see any of the standard textbooks on Kähler geometry (e.g., [Wells73, V.4.1; Griffiths-Harris78, p. 116]).

9.3 THEOREM. *Let X be a smooth proper variety (or compact Kähler manifold) with structure sheaf $\mathcal{O}_X$. Let $\mathbb{C}_X \subset \mathcal{O}_X$ denote the constant sheaf. Then*

$$H^i(X, \mathbb{C}_X) \to H^i(X, \mathcal{O}_X) \quad \textit{is surjective for every } i. \qquad \blacksquare$$

The method of proof of the vanishing theorems relies on an auxiliary construction, which we now introduce.

9.4 Cyclic covers. Let X be an algebraic variety and L a line bundle on X. The total space $T(L)$ of L is again an algebraic variety together with a morphism $p : T(L) \to X$. (I tend to think of L as a coherent sheaf, even if I say line bundle. This is the main reason for the extra notation $T(L)$.)

$$p_*\mathcal{O}_{T(L)} = \sum_{i=0}^{\infty} L^{-i} \quad \text{hence} \quad p_*(p^*L) = \sum_{i=-1}^{\infty} L^{-i}.$$

p^*L has a canonical section y_L corresponding to the section 1 of the summand $L^0 = \mathcal{O}_X$. More geometrically, if $z \in T(L)$ then $y_L(z) = z$ (with suitable indentifications).

Let $t \in \Gamma(X, L)$ be a section. We can also view t as a hypersurface in $T(L)$. Using the above notation it is the set of zeros of the section $y_L - p^*t \in \Gamma(T(L), p^*L)$.

Pick any point $x \in X$ and let L_x be the fiber of L over x. The nth power map

$$\text{mult} : L_x \to L_x^{\otimes n} \quad \text{given by} \quad z \mapsto z^{\otimes n}$$

gives a global morphism $\text{mult} : T(L) \to T(L^{\otimes n})$.

Let s be a section of $L^{\otimes n}$. The choice of s (up to a global section of $\mathcal{O}_X^*$) is equivalent to writing $L^{\otimes n} \cong \mathcal{O}_X(D)$ for some effective divisor D. We write $D = \sum d_i D_i$ if we want to consider the irreducible components of D with their multiplicities.

As before, s can be viewed as a hypersurface in $T(L^{\otimes n})$ given by the equation $y_{L^n} - s = 0$. $\text{mult}^{-1}(s) \subset T(L)$ is called the variety obtained by *taking the nth root of s*. It is sometimes denoted by $X[\sqrt[n]{s}]$. By definition $y_{L^n} = (y_L)^n$. Thus

$$\text{(9.4.1)} \quad [\sqrt[n]{s}] = ((y_L)^n - p^*s = 0) \subset T(L).$$

From this description we obtain the following result, which is frequently taken as the definition of $X[\sqrt[n]{s}]$.

9.5 PROPOSITION.

$$p_*\mathcal{O}_{X[\sqrt{s}]} = \sum_{j=0}^{n-1} L^{-j},$$

and the multiplication is given by the rules

$$L^{-a} \otimes L^{-b} = L^{-a-b} \qquad \text{if } a+b \le n-1, \text{ and}$$

$$L^{-a} \otimes L^{-b} = L^{-a-b} \xrightarrow{\otimes s} L^{-a-b+n} \qquad \text{if } a+b \ge n.$$

Proof. $X[\sqrt{s}]$ is given by the exact sequence

$$0 \to p^*L^{-n} \xrightarrow{y_L^n - p^*s} \mathcal{O}_{T(L)} \to \mathcal{O}_{X[\sqrt{s}]} \to 0.$$

Thus $p_*\mathcal{O}_{X[\sqrt{s}]}$ is the quotient of $p_*\mathcal{O}_{T(L)}$ by the image of

$$y_L^n - p^*s : \sum_{i=n}^{\infty} L^{-i} \to \sum_{i=0}^{\infty} L^{-i}.$$

$y_L^n : L^{-n} \to L^{-n}$ is an isomorphism and $s : L^{-n} \to \mathcal{O}_X$ is the dual of the injection $s : \mathcal{O}_X \to L^n$. Thus if $g \in \Gamma(U, L^{-n})$ is a local section, then $g - sg = (y_L^n - p^*s)g$ is zero in $p_*\mathcal{O}_{X[\sqrt{s}]}$. In down-to-earth terms: replace any section g of L^{-n} by its image sg in $\mathcal{O}_X$. ∎

It is easy to describe $X[\sqrt{s}]$ locally. Pick a local coordinate system (x_i) on $U \subset X$. Then $y = y_L$ and the x_i give a local coordinate system on $p^{-1}(U)$. In these coordinates the nth power morphism becomes $(x_i, y) \mapsto (x_i, y^n)$. Thus $X[\sqrt{s}]$ is locally given by the equation

(9.5.1) $\quad y^n = s(x_1, \ldots, x_n).$

By an easy computation we get the following.

9.6 COROLLARY. *Assume that $x \in X$ is a smooth point. $X[\sqrt{s}]$ is smooth above $x \in X$ iff either $s(x) \neq 0$ or $(s = 0)$ is smooth at x.* ∎

Unfortunately $X[\sqrt{s}]$ is not normal if any of the coefficients in $\sum d_i D_i$ is greater than 1. In order to get its normalization, we need some notation.

9.7 Definition. For a real number x let $\lfloor x \rfloor$ (called "round down") be the largest integer $\le x$, $\lceil x \rceil$ (called "round up") the smallest integer

$\geq x$ and $\{x\} := x - \llcorner x \lrcorner$ (called "fractional part"). If $D = \sum d_i D_i$ is a formal linear combination of distinct irreducible divisors, then set $\llcorner D \lrcorner := \sum \llcorner d_i \lrcorner D_i$, $\ulcorner D \urcorner := \sum \ulcorner d_i \urcorner D_i$ and $\{D\} := \sum \{d_i\} D_i$.

9.8 PROPOSITION. *Let X be a normal variety, Z the normalization of $X[\sqrt[n]{s}]$ and $\bar{p} : Z \to X$ the natural morphism. Then*

$$\bar{p}_* \mathcal{O}_Z = \sum_{j=0}^{n-1} L^{-j}(\llcorner jD/n \lrcorner) = \sum_{j=0}^{n-1} L^{-j}\Big(\sum_i \llcorner jd_i/n \lrcorner D_i\Big).$$

Proof. Fix an nth root of unity ϵ. $y \mapsto \epsilon y$ defines a $\mathbb{Z}_n$ action on $X[\sqrt[n]{s}]$ hence also on $p_* \mathcal{O}_{X[\sqrt[n]{s}]}$. $\mathbb{Z}_n$ acts on the summand L^{-j} via multiplication by ϵ^{-j}. The $\mathbb{Z}_n$ action extends to the normalization Z and thus we have a $\mathbb{Z}_n$ action on

$$(9.8.1) \quad \bar{p}_* \mathcal{O}_Z \supset p_* \mathcal{O}_{X[\sqrt[n]{s}]}.$$

We can decompose $\bar{p}_* \mathcal{O}_Z$ as the sum of eigensheaves

$$\bar{p}_* \mathcal{O}_Z = \sum_{j=0}^{n-1} F_j, \quad \text{and} \quad F_j \supset L^{-j}.$$

$\bar{p}_* \mathcal{O}_Z$ is torsion free, reflexive, and the inclusion (9.8.1) is generically an isomorphism. Thus F_j is a rank 1 torsion-free reflexive sheaf which contains L^{-j}. Hence $F_j = L^{-j}(E_j)$ for some effectve divisor $E_j \subset X$. We need to compute E_j.

Let g be a rational section of L^{-j}. g is a section of $F_j \Leftrightarrow g$ is integral over $\mathcal{O}_X \Leftrightarrow g^n$ is integral over $\mathcal{O}_X \Leftrightarrow g^n \in \mathcal{O}_X$. Therefore E_j is the largest divisor such that

$$F_j^n = (L^{-j}(E_j))^n = L^{-nj}(nE_j) = \mathcal{O}_X(-jD + nE_j)$$

is a subsheaf of $\mathcal{O}_X$. This means that $E_j \leq jD/n$, hence $E_j = \llcorner jD/n \lrcorner$. ∎

9.9 Example. Let us compute explicitly the example where X is an open subset in $\mathbb{C}^m$, $L = \mathcal{O}_X$, and $s = x_1^d$. (Assume that x_1 is not invertible on X.)

$X[\sqrt[n]{s}]$ is defined by the equation $y^n - x_1^d = 0$ in $\mathbb{C}^{m+1}$. n and d may not be relatively prime, so write $n = n'(n,d)$ and $d = d'(n,d)$. $X[\sqrt[n]{s}]$ has (n,d) irreducible components, each isomorphic to $y^{n'} - x_1^{d'} = 0$.

The normalizations of these components are isomorphic to the same open subset in $\mathbb{C}^m$. Explicitly, take coordinates $(z, x_2, \ldots, x_m)$ and look at

$$\mathbb{C}^m \to \mathbb{C}^{m+1} \quad \text{given by}$$
$$(z, x_2, \ldots, x_m) \mapsto (z^{d'}, z^{n'}, x_2, \ldots, x_m).$$

The image is a hypersurface defined by $y^{n'} - x_1^{d'} = 0$ and the map is injective, hence birational.

Thus the normalization of $X[\sqrt{s}]$ is smooth. It has (n, d) connected components. The Galois group $\mathbb{Z}_n$ acts transitively on these components. The subgroup $\mathbb{Z}_{n'} < \mathbb{Z}_n$ acts on each component by

$$(z, x_2, \ldots, x_m) \to (\epsilon^j z, x_2, \ldots, x_m),$$

where ϵ is a primitive n'^{th} root of unity. Thus the $\mathbb{Z}_{n'}$ action is trivial iff $n' = 1$, which is equivalent to $n|d$.

9.10 Definition. Notation as in (9.4). Let $\bar{p} : Z \to X$ be the nomalization of $X[\sqrt{s}]$. Under the $\mathbb{Z}_n$ action $\bar{p}_*\mathbb{C}_Z$ decomposes into eigensheaves $\sum G_j$.

$\bar{p}_*\mathbb{C}_Z$ is naturally a subsheaf of $\bar{p}_*\mathcal{O}_Z$, and we fix the indexing such that G_j is a subsheaf of $F_j = L^{-j}(\llcorner jD/n \lrcorner)$ defined in (9.8).

9.10.1 Remark. In (9.16) we determine the sheaves G_j explicitly. It is somewhat surprising that the explicit forms are of very little help in the applications. Proving the vanishing theorems without knowing the sheaves explicitly saves a little time and makes the general ideas behind the proof trasparent.

9.11 PROPOSITION. *Notation as above.*

(9.11.1) $G_0 \cong \mathbb{C}_X$.

(9.11.2) For every $x \in X$ *there is an open neighborhood* $x \in U_x \subset X$ *such that* $H^i(U_x, G_j|U_x) = 0$ *for* $i > 0$.

(9.11.3) If $U \subset X$ *is connected,* $U \cap D \neq \emptyset$ *and* n *does not divide any of the coefficients of* $D = \sum d_i D_i$, *then* $H^0(U, G_1|U) = 0$.

(9.11.4) If X *is smooth, proper, and Poincaré duality holds for* $\mathbb{C}_Z$ *on* Z, *then the natural pairing*

$$H^i(X, G_j) \times H^{2\dim X - i}(X, G_{\dim X - j}) \to H^{2\dim X}(X, \mathbb{C}_X) = \mathbb{C}$$

is perfect.

Proof. G_0 is the invariant part of $\bar{p}_*\mathbb{C}_Z$, which is $\mathbb{C}_X$.

Choose $U_x \subset X$ such that $V_x = (\bar{p})^{-1}U_x \subset Z$ is contractible. Thus $H^i(V_x, \mathbb{C}_{V_x}) = 0$ for $i > 0$. Since $\bar{p}$ is finite,

$$H^i(W, \bar{p}_*\mathbb{C}_Z|W) = H^i((\bar{p})^{-1}W, \mathbb{C}_Z|(\bar{p})^{-1}W),$$

for every $W \subset X$. In particular for $W = U_x$ we obtain that $H^i(U_x, \bar{p}_*\mathbb{C}_Z|U_x) = 0$. This implies (9.11.2) since $H^i(U_x, G_j|U_x)$ is a direct summand of $H^i(U_x, \bar{p}_*\mathbb{C}_Z|U_x)$.

If U is connected and U intersects D, then we can find a point $x \in U \cap D$, which has a neighborhood where X is smooth and D is defined by a power of a coordinate function. By assumption $n' = n/(n, d) > 1$. By the computations of (9.9) every global section of $G_1|U$ is invariant under the $\mathbb{Z}_{n'}$ action. By definition the $\mathbb{Z}_{n'}$ action on G_1 is nontrivial. Thus $G_1|U$ has no global sections.

Finally (9.11.4) is just the Poincaré duality on Z decomposed into $\mathbb{Z}_n$-eigenspaces. ∎

The following is the most general technical surjectivity result for line bundles that I know. It represents a culmination of the work of several authors: [Tankeev71; Ramanujam72; Miyaoka80; Kawamata82; Viehweg82; Kollár86a,b,87; Esnault-Viehweg86,87].

The rest of this chapter and the next one are devoted to deriving various applications.

9.12 THEOREM. *Let X be a proper variety and L a line bundle on X. Let $L^n \cong \mathcal{O}_X(D)$, where $D = \sum d_i D_i$ is an effective divisor. Assume that $0 < d_i < n$ for every n. Let Z be the normalization of $X[\sqrt{s}]$. Assume furthermore that*

(9.12.1) $H^j(Z, \mathbb{C}_Z) \to H^j(Z, \mathcal{O}_Z)$ is surjective.

Then

(9.12.2) $H^j(X, G_1) \to H^j(X, L^{-1})$ is surjective;

(9.12.3) for any $b_i \geq 0$ the natural map

$$H^j\left(X, L^{-1}\left(-\sum b_i D_i\right)\right) \to H^j(X, L^{-1}) \quad \textit{is surjective.}$$

Proof. As before, let $\bar{p}_*\mathbb{C}_Z = \sum G_k$ and $\bar{p}_*\mathcal{O}_Z = \sum F_k$ be the eigensheaf decompositions. Since $\bar{p}$ is finite, $H^j(Z, \mathbb{C}_Z) = H^j(X, \bar{p}_*\mathbb{C}_Z)$ and similarly for $\mathcal{O}_Z$. By assumption the natural map

$$\sum_k H^j(X, G_k) \to \sum_k H^j(X, F_k)$$

is surjective; thus it is surjective on every $\mathbb{Z}_n$-eigenspace $H^j(X, G_k) \to H^j(X, F_k)$.

$0 < d_i < n$, thus $\llcorner d_i/n \lrcorner = 0$ and $F_1 = L^{-1}(\llcorner D/n \lrcorner) = L^{-1}$. This shows (9.12.2).

9.12.4 CLAIM. *Notation as above. Then G_1 is a subsheaf of $L^{-1}(-\sum b_i D_i)$.*

Proof. Both of these are subsheaves of L^{-1}, so this is a local question. $L^{-1}(-\sum b_i D_i)$ and L^{-1} are equal over $Y - D$. If $U \subset X$ is connected and it intersects D, then by (9.11.3) $H^0(U, G_1|U) = 0$, thus $G_1|U \subset L^{-1}(-\sum b_i D_i)|U$ trivially. ∎

This gives a factorization

$$H^j(X, G_1) \to H^j\left(X, L^{-1}\left(-\sum b_i D_i\right)\right) \to H^j(X, L^{-1}).$$

The composition is surjective by (9.12.2), hence the second arrow is also surjective. ∎

9.12.5 Complement. Let L be a rank 1 reflexive sheaf such that the reflexive hull of $L^{\otimes n}$ is isomorphic to $\mathcal{O}_X(D)$. Then (9.12) is still true if we use L^{-1} to mean $Hom(L, \mathcal{O}_X)$.

9.13 Two proofs of the Kodaira vanishing theorem. Set $m = \dim X$. Kodaira vanishing is equivalent to $H^j(X, L^{-1}) = 0$ for $j < m$ by Serre duality.

Let X be smooth, projective, and L ample. This implies that a large power of L is generated by global sections. Let $L^n = \mathcal{O}_X(D)$, where D is a smooth divisor. By (9.6) $Z = X[\sqrt[n]{s}]$ is smooth. By (9.3) the assumption (9.12.1) is satisfied and we can apply (9.12). Therefore either of the following results implies (9.1).

9.13.1 LEMMA. *Notation as above. Then*
(9.13.1.1) $H^j(X, G_1) = 0$ for $j < m$;
(9.13.1.2) $H^j(X, L^{-1}(-kD)) = 0$ for $j < m$ and $k \gg 1$.

Proof. For an algebraic geometer, the second part is easier. $H^j(X, L^{-1}(-kD))$ is dual to $H^{m-j}(X, K_X \otimes L \otimes \mathcal{O}(kD))$. $\mathcal{O}(D)$ is ample; thus, by the Serre vanishing theorem, $H^{m-j}(X, F \otimes \mathcal{O}(kD)) = 0$ for $k \gg 1$, $j < m$ and for any sheaf F. Set $F = K_X \otimes L$ to obtain (9.13.1.2).

(9.13.1.1) is more topological. By (9.11.4) $H^j(X, G_1)$ is dual to $H^{2m-j}(X, G_{m-1})$. I claim that

$$(9.13.1.3) \quad H^{2m-j}(X, G_1) = H^{2m-j}(X - D, G_1).$$

Indeed, by (9.11.3) the Čech cohomology of G_1 on X and on $X-D$ is the same. The Čech cohomology agrees that the true cohomology for sheaves that have no higher cohomologies in suitable small neighborhoods of any point (cf. [Hartshorne77, Ex. III.4.11]). Thus (9.11.3) implies (9.13.1.3).

There are at least two ways of proving (9.13.1.1) using (9.13.1.3). The more traditional approach is the following.

Let $D_Z = (\bar{p})^{-1}(D)$. $Z - D_Z \to X - D$ is a finite cover and $H^{2m-j}(X - D, G_1)$ is a direct summand of $H^{2m-j}(Z - D_Z, \mathbb{C})$. D_Z is ample hence $Z - D_Z$ is affine. Therefore it has the homotopy type of an m-dimensional CW complex (see, e.g., [Griffiths-Harris78, p. 158]). Thus $H^{2m-j}(Z - D_Z, \mathbb{C}) = 0$ and so $H^{2m-j}(X - D, G_1) = 0$.

Alternatively, one can prove that if G is any constructible sheaf on an m-dimensional affine variety W, then $H^j(W, G) = 0$ for $j > \dim W$. See [Goresky-MacPherson83, 7.2] for the details. ∎

Another application of the same principle is the following.

9.14 THEOREM. [Kollár86a, 2.1] *Let $f : X \to Y$ be a morphism from a smooth proper variety X to a normal variety Y. Let H be an ample line bundle on Y. Then*

$$H^i(Y, H \otimes R^j f_* K_X) = 0 \quad \text{for } i > 0, j \geq 0.$$

9.14.1 Remark. This proof works if X is a complex manifold that is bimeromorphic to a Kähler manifold. The case when X is Moishezon was treated in [Arapura86].

Proof. Pick a general divisor $H^n \cong \mathcal{O}(E)$ such that $D = f^{-1}(E)$ is smooth. (9.12) and Serre duality imply that

$$(9.14.2) \quad H^i(X, f^*H \otimes K_X) \to H^i(X, f^*H^{1+kn} \otimes K_X)$$

is injective for $i \geq 1$.

We prove (9.14) by induction on $\dim Y$, the assertion being evident if $\dim Y = 0$. We have an exact sequence

$$0 \to K_X \otimes f^*H^t \to K_X \otimes f^*H^{t+n} \to K_D \otimes (f^*H^t|D) \to 0.$$

Using induction and the corresponding long cohomology sequence, we obtain that

$$H^i(Y, H^t \otimes R^j f_* K_X) = H^i(Y, H^{t+n} \otimes R^j f_* K_X) \quad \text{for } i \geq 2.$$

By Serre vanishing,

$$H^i(Y, H^{t+kn} \otimes R^j f_* K_X) \quad \text{for } k \gg 1.$$

Thus

$$(9.14.3) \quad H^i(Y, H^t \otimes R^j f_* K_X) = 0 \quad \text{for } t \geq 1 \text{ and } i \geq 2.$$

Once this much of (9.14) is established, the Leray spectral sequence for $f : X \to Y$ abutting to $H^{j+1}(X, K_X \otimes f^* H^t)$ has only two columns, and therefore it degenerates. This means that

$$H^{j+1}(X, K_X \otimes f^* H^t) = H^0(X, H^t \otimes R^{j+1} f_* K_X) + H^1(X, H^t \otimes R^j f_* K_X).$$

Using (9.14.2) this implies that

$$H^1(X, H \otimes R^j f_* K_X) \to H^1(X, H^{1+kn} \otimes R^j f_* K_X)$$

is injective for every k. As before, this implies that $H^1(X, H \otimes R^j f_* K_X) = 0$. ∎

Finally we compute the sheaves G_j.

9.15 Definition. (9.15.1) Let M be a line bundle on a variety Y and fix an isomorphism $L^n \cong \mathcal{O}_Y$. Let $\mathbb{C}[\rho(L)] \subset L$ be the subsheaf (of Abelian groups) consisting of those local sections g such that $g^n \in \Gamma(Y, \mathcal{O}_Y)$ is locally constant. $\mathbb{C}[\rho(L)]$ is a one-dimensional local system that corresponds to a representation $\rho(L) : \pi_1(Y) \to \mathbb{C}^*$. The image of $\rho(L)$ is contained in the nth roots of unity.

(9.15.2) Let M be a line bundle on a variety Y and fix an isomorphism $L^n \cong \mathcal{O}_Y(D)$, where $D = \sum d_i D_i$ is an effective divisor. This gives an isomorphism $(L|Y - D)^n \cong \mathcal{O}_{Y-D}$; thus we obtain a local system $\mathbb{C}[\rho(L, D)] \subset L|Y - D$, where $\rho = \rho(L, D)$ depends on L and D.

9.16 Proposition. *Notation as in (9.8) and (9.15). Let Z be the normalization of $X[\sqrt{s}]$ with natural map $\bar{p} : Z \to X$. Set*

$$L'_j = L^j(-\llcorner jD/n \lrcorner) \quad \textit{and} \quad D'_j = \sum (jd_i - n\llcorner jd_i/n \lrcorner) D_i).$$

Then

$$\bar{p}_* \mathbb{C}_Z = \sum_{j=0}^{n-1} \mathbb{C}[\rho(L'_j, D'_j)^{-1}] \quad \textit{and} \quad \mathbb{C}[\rho(L'_j, D'_j)^{-1}] = G_j.$$

Proof. By (9.10) G_j is a subsheaf of $L^{-j}(\llcorner jD/n \lrcorner)$. Let g be a local section of $L^{-j}(\llcorner jD/n \lrcorner)$. g is a local section of G_j iff g^n is a locally constant section of

$$(L^{-j}(\llcorner jD/n \lrcorner))^n = \mathcal{O}\left(\sum (n \llcorner jd_i/n \lrcorner - jd_i)D_i\right).$$

Thus $G_j = \mathbb{C}[\rho(L'_j, D'_j)^{-1}]$ by definition. ∎

9.17 Remark. There are many other cases where one can compare the cohomologies of a topological and a coherent sheaf to obtain similar vanishing theorems.

For instance, if $\rho : \pi_1(X) \to U(n)$ is a unitary representation, then one can associate to ρ a local system T_ρ and a locally free coherent sheaf $E_\rho := \mathcal{O}_X \otimes_{\mathbb{C}} T_\rho$. There is a natural map $L_\rho \to E_\rho$. By a result of Deligne (see, e.g., [Zucker79]),

$$H^i(X, T_\rho) \to H^i(X, E_\rho) \quad \text{is surjective for every } i.$$

The analog of (9.11) is clear and under the assumptions of (9.12) we obtain that

$$H^j\left(X, E_\rho \otimes L^{-1}\left(-\sum b_i D_i\right)\right) \to H^j(X, E_\rho \otimes L^{-1})$$

is surjective.

A simple application of this result for one-dimensional representations shows that in (9.12) it is sufficient to assume that L^n and $\mathcal{O}_X(D)$ are numerically equivalent.

Similar results can be formulated for arbitrary variations of Hodge structures over a variety; see [Kollár86b; Kollár87b; Saito90,91].

CHAPTER 10

Generalizations of the Kodaira Vanishing Theorem

In the last decade it became increasingly clear that it is frequently very useful to have analogs of (9.1) and (9.14) for nonample line bundles. The general philosophy suggests that similar results should hold if the line bundle is a "small perturbation" of an ample line bundle. Two notions of perturbation emerged, one in algebraic and one in complex differential geometry. First we recall these definitions and then we prove that they are essentially equivalent. We restrict our attention to smooth varieties, though both versions can be treated on singular spaces as well.

10.1 Definition. Let X be a smooth variety and Δ_i distinct irreducible divisors. Let $\Delta = \sum d_i \Delta_i$ be a formal linear combination with $0 \leq d_i \leq 1$ rational.

(10.1.1) Let $f : Y \to X$ be any proper birational morphism. Let $E_j \subset Y$ denote the exceptional divisors and Δ_i' the birational transform (10.3) of Δ_i. Choose N such that Nd_i are all integers and let $h_i = 0$ be local equations for $(Nd_i)\Delta_i$. Then

$$f^* \left(\left(\prod h_i^{-1} \right) (dx_1 \wedge \cdots \wedge dx_n)^{\otimes N} \right)$$

is a rational section of $K_Y^N(\sum (Nd_i)\Delta_i')$, which is a local generator everywhere outside $\cup E_j$. Thus there are integers e_j such that

$$K_Y^N \left(\sum (Nd_i)\Delta_i' \right) = f^* \left(K_X^N \left(\sum (Nd_i)\Delta_i \right) \right) \otimes \mathcal{O}_Y \left(\sum e_j E_j \right).$$

(10.1.2) Dividing by N we formally write this as

$$K_Y \left(\sum d_i \Delta_i' \right) \equiv f^* \left(K_X + \sum d_i \Delta_i \right) + \sum a(E_j, \Delta) E_j.$$

(10.1.3) The rational number $a(E_j, \Delta) = e_j / N$ is called the *discrepancy* of E_j with respect to (X, Δ).

(10.1.4) If $f' : Y' \to Y \to X$ is another proper birational morphism and E_j' is the birational transform of E_j on Y', then $a(E_j, \Delta) = a(E_j', \Delta)$. This is the reason for suppressing Y in the notation.

(10.1.5) We say that the pair $(X, \sum d_i\Delta_i)$ is *klt* (which stands for *Kawamata log terminal*) if $d_i < 1$ for every i and $a(E, \Delta) > -1$ for every choice of $f : Y \to X$ and for every exceptional divisor $E \subset Y$. (This notion is called "log terminal" in [KaMaMa87].)

(10.1.6) $(X, \sum d_i\Delta_i)$ is called *log canonical* (abbreviated as *lc*) if $a(E, \Delta) \geq -1$ for every choice of $f : Y \to X$ and for every exceptional divisor $E \subset Y$.

10.2 Examples. (10.2.1) Let $\Delta^j = \sum d_i^j\Delta_i$, $j = 1, 2$. If $d_i^1 \leq d_i^2$ for every i and (X, Δ^2) is klt (resp. lc) then (X, Δ^1) is klt (resp. lc).

If $d_i^1 < d_i^2$ for every i and (X, Δ^2) is lc then (X, Δ^1) is klt. (The latter needs the observation that if $f(E_j) \not\subset \cup\Delta_i$ then $a(E_j, \Delta) \geq 0$.)

(10.2.2) Let $\{x_i\}$ be a local coordinate system and $\Delta_i = (x_i = 0)$. Then $(X, \sum d_i\Delta_i)$ is klt iff $d_i < 1$ for every i. Indeed, by (10.2.1) it is sufficient to prove that $(X, \sum \Delta_i)$ is lc.

Pick a morphism $f : Y \to X$ and let $y \in E_j$ be a general point. Pick a local coordinate system $\{y_k\}$ such that $E_j = (y_1 = 0)$. We can write $f^*x_i = y_1^{a_i}u_i$, where u_i is a unit at y. Thus

$$f^*\frac{dx_i}{x_i} = a_i\frac{dy_1}{y_1} + \omega_i, \quad \text{where } \omega_i \text{ is regular at } y.$$

Therefore

$$f^*((x_1 \cdots x_n)^{-1}dx_1 \wedge \cdots \wedge dx_n)$$

has at most a simple pole along E_j.

10.3 Definition. Let X be a variety and $D \subset X$ a divisor. Let $f : Y \to X$ be a proper birational morphism. Let $E \subset Y$ denote the exceptional divisor of f and $D_Y \subset Y$ the *birational transform* of D. (I.e., if f^{-1} is the inverse of f (defined on an open set $X^0 \subset X$), then D_Y is the closure of $f^{-1}(D \cap X^0)$.) (This notion is also called "proper" or "strict transform.")

We say that $f : Y \to X$ is a *log resolution* of (X, D) if Y is smooth and $E + D_Y$ is a divisor with normal crossings only.

The following result gives a practical way of deciding if a given pair is klt or lc:

10.4 LEMMA. *Notation as above. Let $f : Y \to X$ be a log resolution of (X, Δ). (X, Δ) is klt (resp. lc) iff $a(E, \Delta) > -1$ (resp. ≥ -1) for every exceptional divisor E of $f : Y \to X$. (In the klt case we of course need to assume in addition that $d_i < 1$.)*

Proof. Let $f_1 : Y_1 \to X$ be any other proper birational map. Let $f_2 : Y_2 \to X$ be the normalization of $Y_1 \times_X Y$. By factoring $Y_2 \to X$ as $Y_2 \to Y \to X$ and using (10.2.2) we obtain that $f_2 : Y_2 \to X$ satisfies the condition (10.1.5–6). By (10.4) we are done. ∎

10.5 Singular metrics on line bundles. Let L be a line bundle on a complex manifold M. A *singular Hermitian metric* $\| \ \|$ on L is a Hermitian metric on $L|M - Z$ (where Z is a measure zero set) such that if $U \subset M$ is any open subset, $u : L|U \cong U \times \mathbb{C}$ a local trivialization and f a local generating section over U, then

$$\|f\| = |u(f)| \cdot e^{-\phi},$$

where $| \ |$ is the usual absolute value on $\mathbb{C}$ and $\phi \in L^1_{loc}(U)$. (The latter assumption is there for technical resons and we do not use it explicitly. It assures that $\partial\bar{\partial}\phi$ exists as a current on M. Thus we can talk about the curvature form of L, at least as a current on M.)

We say that the metric is L^p on M if $e^{-\phi}$ is locally L^p for every point. (This is clearly independent of the local trivializations.)

10.6 Examples. (10.6.1) Let D be a divisor and $L = \mathcal{O}_X(D)$. L has a natural section f coming from the constant section 1 of $\mathcal{O}_X$. A natural choice of the metric on L is to set $\|f\| = 1$ everywhere. This metric is singular along D. If h is a local equation of D at a point $x \in D$, then $h^{-1}f$ is a local generating section of L at x and

$$\|h^{-1}f\| = e^{-\log|h|}.$$

(10.6.2) Let L be a line bundle on X. Assume that $L^n \cong M(D)$ for some line bundle M and effective divisor D. Let $\| \ \|_M$ be a continuous Hermitian metric on M. As before, this gives a singular metric on L^n. Take n^{th} root to get a metric $\| \ \|_L$ on L.

Let f be a local section of L at a point $x \in D$ and let h be a local equation for D at x. hf^n is a local generator of M and

$$\|f\|_L = (\|hf^n\|_M)^{1/n} e^{-log|h|/n}.$$

The first factor on the right is continuous and positive. Thus $\| \ \|_L$ is L^p iff the exponential factor

$$e^{-log|h|/n} = |h|^{1/n} \quad \text{is } L^p.$$

(10.6.3) Assume that locally at x we can write $D = n\sum d_i\Delta_i$, where $\Delta_i = (x_i = 0)$ for a local coordinate system. Then $\| \|_L$ is L^2 at x iff

$$\prod |x_i|^{-d_i} \quad \text{is } L^2 \text{ near } x,$$

which is equivalent to $d_i < 1$ for every i.

More generally we have the following.

10.7 PROPOSITION. *Let X be a smooth manifold and D a divisor on X. Let L be a line bundle on X and assume that $L^n = M(D)$ for some line bundle M. Set $\Delta = D/n$. Let $\| \|_L$ be the singular metric constructed on L as in (10.6.2). Then*

$$\| \|_L \quad \text{is } L^2 \quad \Leftrightarrow \quad (X, \Delta) \quad \text{is klt.}$$

Proof. Let $f : Y \to X$ be a log resolution of (X, D). Both properties are local in X, so pick a point $x \in D$ and fix a local coordinate system $\{x_i\}$. Let h be a local equation for D. Set $\omega_x = dx_1 \wedge \cdots \wedge dx_k$. $\| \|_L$ is L^2 iff

$$\int |h|^{-2/n} \omega_x \wedge \bar{\omega}_x < \infty. \tag{10.7.1}$$

This is equivalent to saying that $h^{-1}\omega_x^n$ is $L^{2/n}$. The advantage of putting ω_x in is that in this form the condition is invariant under pullbacks. Thus (10.7.1) is equivalent to $f^*(h^{-1}\omega_x^n)$ being $L^{2/n}$ on Y.

This is a local condition on Y, so pick a point and a local coordinate system $\{z_i\}$ such that every component of $E + D_Y$ is defined by a local equation $z_i = 0$ for some i. Set $\omega_z = dz_1 \wedge \cdots \wedge dz_k$. We can write

$$f^*(h^{-1}\omega_x^n) = \omega_z^n \prod z_i^{a_i}.$$

As in (10.6.3), $f^*(h^{-1}\omega_x^n)$ is $L^{2/n}$ iff $a_i > -n$ for every i. Comparing with (10.1.1) we see that $a_i = na(E_i, \Delta)$ if E_i is locally defined by $z_i = 0$, and $a_j = -nd_j$ if Δ'_j is locally defined by $z_j = 0$.

Thus $a_i > -n$ for every i is equivalent to (X, Δ) being klt. ∎

We are ready to formulate the pair of vanishing theorems about perturbations of ample line bundles.

10.8 THEOREM. [Kawamata82; Viehweg82] *Let X be a smooth projective variety and L a line bundle on X. Assume that we can write $L \equiv M + \Delta$ where M is nef and big and (X, Δ) is klt. Then*

$$H^i(X, K_X \otimes L) = 0 \quad \text{for } i > 0.$$

This is proved in (10.14). For the proof of the analytic version we refer to the original papers. Some details are discussed in chapter 11.

10.9 THEOREM. [Demailly82,92; Nadel90b] *Let (X, ω) be a compact Kähler manifold and L a line bundle on X. Let $\| \ \|$ be a singular metric on L such that its curvature form satisfies*

$$\frac{i}{\pi}\partial\bar{\partial}\log\|f\| \geq \epsilon\omega \quad \text{for some } \epsilon > 0.$$

(In most applications the metric is C^2 outside a subvariety $D \subset X$, and the above condition should then hold on $X - D$.) Assume furthermore that $\| \ \|$ is L^2. Then

$$H^i(X, K_X \otimes L) = 0 \quad \text{for } i > 0. \quad \blacksquare$$

The proof of (10.8) runs along the lines of (9.13), but the required cyclic covers are not smooth anymore. The next two results establish their necessary properties.

10.10 PROPOSITION. *Let X be a smooth variety and L a line bundle on X. Let s be a section of L^n such that $(s = 0)$ is a divisor with normal crossings only. Let Z be the normalization of $X[\sqrt{s}]$. Then Z has only quotient singularities.*

Proof. This is a local question. Choose a local coordinate system $\{x_i\}$ such that $D = (\prod x_i^{a_i} = 0)$. Let $U \subset X$ be defined by the inequalities $|x_i| \leq 1$. By (9.5.1) $U[\sqrt{s}]$ is given as $y^n = \prod x_i^{a_i}$.

Let $d = (n, a_1, \ldots, a_k)$. If ϵ is a primitive d^{th} root of unity, then

$$y^n - \prod x_i^{a_i} = \prod_{j=1}^{d}\left(y^{n/d} - \epsilon^j \prod x_i^{a_i/d}\right).$$

Thus after normalization we obtain d different branches, each isomorphic to the normalization of $(y^{n/d} - \prod x_i^{a_i/d} = 0)$. Changing notation we may assume that $(n, a_1, \ldots, a_k) = 1$. Let D^k be the k-dimensional polydisc with coordinates z_i and define a morphism

$$q : D^k \to U[\sqrt{s}] \quad \text{by} \quad x_i = z_i^n, \ y = \prod z_i^{a_i}.$$

Let $G \subset \mathbb{Z}_n^k$ be the subgroup of those elements (u_i) such that $\sum a_i u_i \equiv 0 \mod n$. Then $|G| = n^{k-1}$. q is G-invariant, hence we obtain a morphism $\pi : D^k/G \to U[\sqrt{s}]$. We prove that π is just normalization, which follows once we show that it has degree 1.

$p : U[\sqrt[n]{s}] \to U$ has degree n and $p \circ q : D^k \to U$ has degree n^k. Thus p has degree n^{k-1}. This is the same as the degree of $D^k \to D/G$, thus $D^k/G \to U[\sqrt[n]{s}]$ has degree 1. ∎

10.10.1 Remark. The above representation of $U[\sqrt[n]{s}]$ as a quotient is not necessarily the most economical. If G contains an element of the form $(a, 0, \ldots, 0)$, then we can take the corresponding quotient of D^k which is isomorphic to D^k. This way we get that

$$\bar{U}[\sqrt[n]{s}] \cong D^k/H \quad \text{where}$$
$$H = \ker\left[\sum \mathbb{Z}_{n/(n,a_i)} \xrightarrow{\{u_i\} \mapsto \sum a_i u_i} \mathbb{Z}_n\right].$$

Here $\mathbb{Z}_{n/(n,a_i)}$ acts on $\mathbb{C}^k$ via multiplication by an $(n/(n,a_i))^{th}$ root of unity on the ith coordinate.

10.11 THEOREM. *Let Z be a projective variety with quotient singularities only. Then $H^i(Z, \mathbb{C}_Z) \to H^i(Z, \mathcal{O}_Z)$ is surjective for every i.*

Proof. This can be done in several ways. The easiest is to notice that the usual proof for manifolds works with essentially no changes. We should still view Z as being patched together from smooth coordinate charts, but instead of allowing patching data between different charts only, we admit patching data between a chart and itself, corresponding to the local group action. Once the conceptual difficulties are behind, the proof is really the same.

For other, more conventional proofs, see [Steenbrink77; Danilov78]. ∎

The following result is a direct consequence of (9.12), (9.17), (10.10), and (10.11).

10.12 THEOREM. *Let X be a smooth projective variety (or Kähler manifold) and L a line bundle on X. Let $\sum d_i D_i$ be a divisor with normal crossings only such that $L^n \equiv \mathcal{O}_X(\sum d_i D_i)$ and $0 < d_i < n$ for every i. Then for every $\{b_i \geq 0\}$ and for every j*

$$H^j\left(X, L^{-1}\left(-\sum b_i D_i\right)\right) \to H^j(X, L^{-1}) \quad \textit{is surjective.}$$

By Serre duality this is equivalent to

$$H^j(X, K_X \otimes L) \to H^j\left(X, K_X \otimes L\left(\sum b_i D_i\right)\right) \quad \textit{is injective.} \quad ∎$$

The following is a very general injectivity theorem. It was proved in [Tankeev71] for the case when $j = 1$ and D is a general member of $|kL|$.

The general case for $\Delta = 0$ appears in [Kollár86a]. The $\Delta \neq 0$ case is in [Esnault-Viehweg87].

10.13 THEOREM. *Let $f : X \to Y$ be a surjective morphism from a smooth projective variety (or a compact Kähler manifold) X to a normal variety Y. Let L be a line bundle on X and D an effective divisor on X such that $f(D) \neq Y$. Assume that $L \equiv f^*M + \Delta$, where M is a nef and big $\mathbb{Q}$-divisor on Y and (X, Δ) is klt. Then*

$$H^j(X, K_X \otimes L) \to H^j(X, K_X \otimes L(D)) \quad \text{is injective.}$$

Before proving the theorem, let us derive some corollaries.

10.14 Proof of (10.8). Choose $Y = X$ and let f be the identity. Let D be an ample divisor such that $H^j(X, K_X \otimes L(D)) = 0$ for $j > 0$ (this is possible by Serre vanishing). $D \neq Y$, so by (10.13) $H^j(X, K_X \otimes L) = 0$. ∎

10.15 COROLLARY. [Kollár86a, Esnault-Viehweg87] *Let $f : X \to Y$ be a surjective morphism from a smooth projective variety (or a compact Kähler manifold) X to a normal variety Y. Let L be a line bundle on X such that $L \equiv f^*M + \Delta$, where M is a $\mathbb{Q}$-divisor on Y and (X, Δ) is klt. Then*

(10.15.1) $R^j f_*(K_X \otimes L)$ is torsion free for $j \geq 0$.

(10.15.2) Assume in addition that M is nef and big. Then $H^i(Y, R^j f_*(K_X \otimes L)) = 0$ for $i > 0, j \geq 0$.

Proof. The proof of (10.15.2) follows the arguments of (9.14). Choose H ample and substitute $K_X \otimes L$ for K_X in the proof. We obtain the analog of (9.14.3):

$$H^i(Y, H^t \otimes R^j f_*(K_X \otimes L)) = 0 \quad \text{for } t \geq 0 \text{ and } i \geq 2.$$

The rest of the argument is again the same, proving (10.15.2).

(10.15.2) implies that if H is sufficiently ample on Y then

$$H^j(X, K_X \otimes L \otimes f^*H) = H^0(Y, H \otimes R^j f_*(K_X \otimes L))$$

for any M. Thus by (10.13) if D is any divisor on Y then the natural map

$$(10.15.3) \qquad \begin{aligned} &H^0(Y, H \otimes R^j f_*(K_X \otimes L)) \\ &\quad \to H^0(Y, H(D) \otimes R^j f_*(K_X \otimes L)) \end{aligned}$$

is injective. Assume that $F \subset R^j f_*(K_X \otimes L)$ is a torsion subsheaf. Choose H and D such that $H \otimes F$ is generated by global sections and multiplication by the equation of D kills F. Then the map (10.15.3) kills $H^0(X, H \otimes F)$, which implies that $F = 0$. ∎

10.15.4 Remarks. (10.15.4.1) By suitable base-change tricks the above proof can be made to work if Y is only Moishezon.

(10.15.4.2) The more general case of Kähler morphisms is considered in [Saito90,91].

(10.15.4.3) [Nakamura75] gives an example of a smooth family $\{X_t\}$ of compact complex threefolds where $h^i(X_t, K_{X_t})$ jumps. Thus some sort of Kähler assumption is essential.

10.16 COROLLARY. [Grauert-Riemenschneider70] *Let $f : X \to Y$ be a birational morphism of projective varieties, X smooth. Then $R^j f_* K_X = 0$ for $j > 0$.*

More generally, if $L \equiv N + \Delta$ where N is an f-nef $\mathbb{Q}$-divisor and (X, Δ) is klt, then $R^j f_*(K_X \otimes L) = 0$ for $j > 0$.

Proof. Choose H sufficiently ample on Y such that $H \otimes R^j f_*(K_X \otimes L)$ is generated by global sections and

$$H^j(X, K_X \otimes L \otimes f^*H) = H^0(Y, H \otimes R^j f_*(K_X \otimes L)).$$

If in addition $N + f^*H$ is nef and big, then the group on the left-hand side vanishes for $j > 0$ by (10.8), hence $R^j f_*(K_X \otimes L) = 0$ for $j > 0$.

Unfortunately, one cannot always make $N + f^*H$ nef and big. Instead we try to achieve that $N + f^*H - \Delta'$ is ample where Δ' is a very small effective divisor such that $(X, \Delta + \Delta')$ is still klt. The rest of the proof then works.

Let A be an ample effective divisor on X. Choose H_0 ample on Y such that $f^*H_0 = A + A'$ where A' is effective. Then $N + \epsilon A$ is f-ample and $N + \epsilon A + (m - \epsilon)H_0$ is ample for $m \gg 1$. Set $H = mH_0$. Then

$$N + f^*H - \epsilon A' \equiv N + (\epsilon A + (m - \epsilon)f^*H_0)$$

and $(X, \Delta + \epsilon A')$ is klt for $0 < \epsilon \ll 1$. ∎

10.17 Proof of (10.13). Since M is nef and big, one can find $n \gg 1$ such that $f^*(nM) \sim D + D'$ for some effective divisor D'. Thus

$$nL \sim D + D' + n\Delta.$$

There are two reasons why (10.12) cannot be applied directly. First, the divisor $D + D' + n\Delta$ need not have normal crossing. Second, it can have too large coefficients.

10.17.1 Special case. Assume first that M is ample and $D + D' + n\Delta$ is a divisor with normal crossings. For $k \gg 1$ we can choose a smooth divisor $M_k \in |kf^*M|$ such that $D + D' + \Delta + M_k$ has normal crossings. Thus we can write

$$(n+m)L \equiv D + D' + (n+m)\Delta + \frac{m}{k}M_k.$$

For $m \gg 1$ every coefficient on the right-hand side is less than $n+m$. Thus (10.12) applies to show that (10.13) holds in this special case.

The rest of the proof consists of steps that reduce the general case to the special one.

10.17.2 Reduction to M ample. This is quite easy and general. Let H be ample on Y. By (0.4.2) we can write $nM \sim H + E$ for $n \gg 1$ and E effective. Thus

$$M \equiv \frac{1}{n+m}(mM + H) + \frac{1}{n+m}E \quad \text{and } mM + H \text{ is ample.}$$

Thus we can write

$$L \equiv f^*M + \Delta \equiv f^*\left(\frac{1}{n+m}(mM+H)\right) + \left(\Delta + \frac{1}{n+m}f^*E\right)$$

and $(X, \Delta + \frac{1}{n+m}f^*E)$ is klt for $m \gg 1$.

$D + D' + \Delta$ can be brought to normal crossing after a sequence of blowing ups. In order to control the change in the cohomology groups in the process, we need another special case of (10.13):

10.17.3 CLAIM. *Notation and assumptions as in (10.13). Assume in addition that f is birational, Y is smooth, and the union of the exceptional divisor of f and of Δ has normal crossings only. Then*

(10.17.3.1) $R^i f_*(K_X \otimes L) = 0$ for $i > 0$;

(10.17.3.2) if M is ample on Y then $H^i(X, K_X \otimes L) = 0$.

Proof. Assume first that M is ample. Let $E = \sum e_i E_i$ be an effective exceptional divisor such that $nf^*M - E$ is ample on X for $n \gg 1$. (For instance, pick an ample divisor A on X and let $E = f^*f_*A - A$.)

Let D be a general member of $|m(nf^*M - E)|$ for $m \gg 1$. Then

$$\frac{1}{m}D + E + n\Delta \equiv nL$$

has normal crossings and all coefficients are below n for $n \gg 1$. Thus

$$H^i(X, K_X \otimes L) \to H^i(X, K_X \otimes L(kD))$$

is injective by (10.12). By Serre vanishing we get (10.17.3.2).

Choose H sufficiently ample on Y such that

$$H^j(Y, H \otimes R^i f_*(K_X \otimes L)) = 0 \quad \text{for } j > 0, i \geq 0.$$

With this choice of H,

$$H^0(Y, H \otimes R^i f_*(K_X \otimes L)) = H^i(X, K_X \otimes L \otimes f^*H).$$

The latter group is zero for $i > 0$ by the first step. We may assume that $H \otimes R^i f_*(K_X \otimes L)$ is generated by global sections, thus it is zero. ∎

10.17.4 The general case. By (10.17.2) we may assume that M is ample. Let $p : Z \to X$ be a log resolution of $(X, D + D' + \Delta)$.

We use (10.17.3) to reduce (10.13) to a problem on Z with carefully chosen L_Z, D_Z, and Δ'_Z. By (10.1.2) we have

$$K_Z + \Delta_Z \equiv p^*(K_X + \Delta) + \sum a(E_i, \Delta)E_i.$$

By assumption $a(E_i, \Delta) > -1$ for every i. Rewrite the above equality as

$$K_Z + \Delta_Z + \sum\{-a(E_i, \Delta)\}E_i \equiv p^*(K_X + \Delta) + \sum \ulcorner a(E_i, \Delta)\urcorner E_i.$$

Set

$$\begin{aligned} &\Delta'_Z = \Delta_Z + \sum\{-a(E_i, \Delta)\}E_i, \quad \text{and} \\ &L_Z = p^*L(-\llcorner p^*(\Delta)\lrcorner) \equiv (f \circ p)^*M + \Delta'_Z. \end{aligned} \tag{10.17.4.1}$$

Δ'_Z is an effective normal crossing divisor all of whose coefficients are less than 1.

$$\begin{aligned} &K_Z \otimes L_Z = p^*(K_X \otimes L) \otimes \mathcal{O}_Z(\textstyle\sum \ulcorner a(E_i, \Delta)\urcorner E_i) \quad \text{and,} \\ &K_Z \otimes L_Z \equiv K_Z + (f \circ p)^*M + \Delta'_Z. \end{aligned}$$

This implies that

$$p_*(K_Z \otimes L_Z) = K_X \otimes L \quad \text{and}$$
$$p_*(K_Z \otimes L_Z(p^*D)) = K_X \otimes L(D).$$

Applying (10.17.3.1) to $p : Z \to X$ and using (10.17.4.1) we obtain that

$$R^j p_*(K_Z \otimes L_Z) = R^j p_*(K_Z \otimes L_Z(p^*D)) = 0 \quad \text{for } j > 0.$$

Thus

$$H^j(Z, K_Z \otimes L_Z) = H^j(X, K_X \otimes L), \quad \text{and}$$
$$H^j(Z, K_Z \otimes L_Z(p^*D)) = H^j(X, K_X \otimes L(D)).$$

Therefore it is sufficient to prove injectivity for

(10.17.4.2) $\quad H^j(Z, K_Z \otimes L_Z) \to H^j(Z, K_Z \otimes L_Z(p^*D)).$

By construction $nL \sim p^*(D + D') + n\Delta'_Z$, and the right-hand side is a divisor with normal crossings. Thus (10.17.4.2) is injective by (10.17.1). ∎

For the sake of completeness I indicate briefly how (10.13) and its corollaries can be generalized to singular varieties. The crucial step is the following.

10.18 PROPOSITION. *Let X be a normal variety, B a $\mathbb{Q}$-Cartier Weil divisor, Δ a $\mathbb{Q}$-Weil divisor, and N a $\mathbb{Q}$-Cartier $\mathbb{Q}$-divisor. Assume that (X, Δ) is klt and $B \equiv K_X + \Delta + N$. Let $p : Z \to X$ be a log resolution of $(X, \Delta + B)$.*

Then there is a Cartier divisor B_Z on Z and a normal crossing divisor Δ_Z such that

(10.18.1) (Z, Δ_Z) is klt and $p_*\Delta_Z = \Delta$;
(10.18.2) $B_Z \equiv K_Z + \Delta_Z + p^*N$;
(10.18.3) $p_*\mathcal{O}_Z(B_Z) = \mathcal{O}_X(B)$ and $R^i p_*\mathcal{O}_Z(B_Z) = 0$ for $i > 0$.

Proof. Let $E_i \subset Z$ be the exceptional divisors of p. Define the numbers $a_i > -1$, $d_i \geq 0$ by the formulas

$$K_Z + p_*^{-1}\Delta \equiv p^*(K_X + \Delta) + \sum a_i E_i, \quad \text{and}$$
$$p^*B \equiv \llcorner p^*B \lrcorner + \sum d_i E_i.$$

Set

$$B_Z := \llcorner p^*B \lrcorner + \sum \ulcorner a_i + d_i \urcorner E_i, \quad \text{and}$$
$$\Delta_Z := p_*^{-1}\Delta + \sum \{-a_i - d_i\} E_i.$$

It is easy to see that all the requirements are satisfied. ∎

Using the arguments of (10.17.4) we see that (10.13) and (10.18) imply the following more general result.

10.19 THEOREM. *Let $f : X \to Y$ be a surjective morphism between normal and proper varieties. Let N, N' be rank 1, reflexive, torsion-free sheaves on X. Assume that $N \equiv K_X + \Delta + f^*M$, where M is a $\mathbb{Q}$-Cartier $\mathbb{Q}$-divisor on Y and (X, Δ) is klt. Then*

(10.19.1) $R^j f_* N$ is torsion free for $j \geq 0$.

(10.19.2) Assume in addition that M is nef and big. Then

$$H^i(Y, R^j f_* N) = 0 \quad \textit{for } i > 0, j \geq 0.$$

(10.19.3) Assume that M is nef and big and let D be an effective Weil divisor on X such that $f(D) \neq Y$. Then

$$H^j(X, N) \to H^j(X, N(D)) \quad \textit{is injective for } j \geq 0.$$

(10.19.4) If f is generically finite and $N' \equiv K_X + \Delta + F$ where F is a $\mathbb{Q}$-Cartier $\mathbb{Q}$-divisor on X which is f-nef, then $R^j f_* N' = 0$ for $j > 0$. ∎

CHAPTER 11

Vanishing of L^2-Cohomologies

In this chapter I will give a quick review of the results of [Demailly82,92] about vanishing theorems for line bundles with singular Hermitian metrics.

11.1 Definition-Proposition. (11.1.1) Let X be a complex manifold. By a *degenerate Kähler form* we mean a closed positive Kähler form ω defined on a Zariski dense open set $V \subset X$ such that if $(z_1, \dots, z_n)$ is any local coordinate chart on X, then

$$\omega = \sum g_{ij} dz_i \wedge d\bar{z}_j$$

and the g_{ij} are bounded. We denote by $dV(\omega) = i^{n(n-1)}\omega^n$ the corresponding *degenerate volume form*. (The definition is dictated by immediate needs rather than by any theoretical considerations.)

(11.1.2) Let X be a complex manifold with a degenerate Kähler form ω. Let E be a vector bundle on X. By a *singular Hermitian metric* on E we mean a C^∞ Hermitian metric $h(\ ,\ )$ defined over a Zariski dense open set $V \subset X$.

We say that h is L^2 with respect to ω if $h(s_1, s_2)dV(\omega)$ is locally integrable on B for any local coordinate chart $B \subset X$ and for any pair of bounded and measurable (with respect to some nonsingular Hermitian metric) local sections s_1, s_2 of E. (One could argue that under the above assumptions h should be called L^1. h is a quadratic tensor, thus the above definition makes some sense. It is consistent with (10.5).)

(11.1.3) If a Hermitian metric h is L^2 with respect to a Kähler form, then it is also L^2 with respect to any degenerate Kähler form. The converse is false.

(11.1.4) Let ω be a degenerate Kähler form on X which is a Kähler form on $V \subset X$. If $f : Y \to X$ is a generically finite morphism and $f(Y)$ intersects V, then $f^*\omega$ is again a degenerate Kähler form.

(11.1.5) If $f : Y \to X$ is a generically finite and dominant morphism, then f^*h is again a singular L^2 metric with respect to $f^*\omega$. ∎

The following result shows that the L^2 condition on sections of a vector bundle is largely independent of the choice of the metrics.

11.2 PROPOSITION. *Let X be a compact complex manifold and E a vector bundle on X. Choose degenerate Kähler forms ω_i on X and singular Hermitian metrics h_i on E such that h_i is L^2 with respect to ω_i $(i = 1, 2)$. We denote by $\tilde{}$ the corresponding objects on the universal cover of X. (We could work on any cover.)*

There is a constant c such that for any a holomorphic section f of $\tilde{E}$

$$\int_{\tilde{X}} h_1(f, \bar{f})dV(\omega_1) \leq c \int_{\tilde{X}} h_2(f, \bar{f})dV(\omega_2).$$

In particular, the space of holomorphic L^2 sections does not depend on the choice of the degenerate Kähler form and the singular Hermitian metric.

Proof. During the proof c stands for a variable positive constant.

Let $Z \subset X$ be the union of the degeneracy loci and U a small open neighborhood of Z. $X - U$ is compact, ω_i and h_i are continuous on $X - U$, thus they are bounded by constant multiples of each other. Therefore,

$$\int_{\tilde{X}-\tilde{U}} h_1(f, \bar{f})dV(\omega_1) \leq c \int_{\tilde{X}-\tilde{U}} h_2(f, \bar{f})dV(\omega_2),$$

and it is sufficient to prove that

$$\int_{\tilde{X}} h_1(f, \bar{f})dV(\omega_1) \leq c \int_{\tilde{X}-\tilde{U}} h_1(f, \bar{f})dV(\omega_1). \tag{11.2.1}$$

This concerns only one metric, so we drop the subscript. By resolution of singularities we may assume that Z is contained in a divisor with normal crossings. Thus (11.2) is implied by the following local statement. ($D(r)$ denotes the n-dimensional polydisc of radius r.)

11.2.2 LEMMA. *Let ω be a degenerate Kähler form on $D(1+\epsilon)$ and E a vector bundle over $D(1+\epsilon)$ with an L^2 metric with respect to ω. Assume that ω and h are both nondegenerate outside the coordinate hyperplanes. Let $T \subset D(1+\epsilon)$ be the "fat torus" $T = \{(z_1, \dots, z_n) | 1/2 \leq |z_i| \leq 1\}$.*

There is a constant $c > 0$ such that for any holomorphic section s of E

$$\int_{D(1/2)} |h(s, \bar{s})| dV(\omega) \leq c \int_T |h(s, \bar{s})| dV(\omega).$$

Proof. Fix a trivialization of E given by global sections e_i of E and set $h_{ij} = h(e_i, e_j)$. On T the Hermitian metric h and the diagonal metric

$(e_\iota, e_j) = \delta_{\iota j}$ are mutually bounded by constant multiples of each other, thus it is sufficient to prove that

$$\int_{D(1/2)} |h_{\iota j} f_\iota \bar{f}_j| dV(\omega) \le c \int_T |f_\iota|^2 + |f_j|^2 dV(\omega).$$

Write $dV(\omega) = \phi dV$, where dV is the standard volume form $i^{n(n-2)} dz_1 \wedge d\bar{z}_1 \wedge \cdots \wedge dz_n \wedge d\bar{z}_n$. Using that $|\phi|$ is bonded from above and from below on T, we are reduced to the inequality

$$\int_{D(1/2)} |h_{\iota j}| \cdot |f_\iota \bar{f}_j| dV(\omega) \le c \int_T |f_\iota|^2 + |f_j|^2 dV.$$

The f_ι are holomorphic, $|f(\mathbf{z})| \le c \int_T |f(\mathbf{x})| dV$ for $\mathbf{z} \in D(1/2)$. Thus

$$\begin{aligned}
&\int_{D(1/2)} |h_{\iota j}| \cdot |f_\iota \bar{f}_j| dV(\omega) \\
&\quad \le c \int_{D(1/2)} |h_{\iota j}| dV(\omega) \int_T |f_\iota(\mathbf{x})| dV \int_T |f_j(\mathbf{x})| dV \\
&\quad \le c \int_T |f_\iota|^2 + |f_j|^2 dV.
\end{aligned}$$

Here we used that $\int_{D(1/2)} |h_{\iota j}| dV(\omega) < \infty$ which is exactly the L^2 assumption on h with respect to ω. ∎

11.2.3 COROLLARY. *Let X be a compact complex manifold and $u : \tilde{X} \to X$ its universal cover. Let L_1, L_2 be line bundles on X with Hermitian metrics h_ι. Let $s \in H^0(X, L_2)$ be a nonzero section. Then multiplication by $u^* s$:*

$$m(u^* s) : (L^2 \cap H)(\tilde{X}, u^* L_1) \to (L^2 \cap H)(\tilde{X}, u^*(L_1 \otimes L_2))$$

is injective with closed range.

Proof. Define a singular Hermitian metric h on $u^* L_1$ by $h(f, \bar{f}) = (h_1 \otimes h_2)(f \otimes s, \bar{f} \otimes \bar{s})$. By (11.2) h and h_1 induce quasi-isometric metrics on $(L^2 \cap H)(\tilde{X}, u^* L_1)$. Thus if $m(u^* s)(f_\iota)$ is a Cauchy sequence then so is f_ι. ∎

11.3 COROLLARY. *Let $f : X_1 \to X_2$ be a proper and bimeromorphic morphism between compact complex manifolds. Let E_ι be a vector bundle on X_ι. Assume that we have a map $j : f^* E_2 \to E_1$ which is an isomorphism outside the exceptional set $E \subset X_2$ of f.*

Choose degenerate Kähler forms ω_i on X_i and singular Hermitian metrics h_i on E_i such that h_i is L^2 with respect to ω_i $(i = 1, 2)$. There is a natural quasi-isometry

$$\tilde{f}^* : H^0_{(2)}(\tilde{X}_2, \tilde{\omega}_2, \tilde{E}_2, \tilde{h}_2) \cong H^0_{(2)}(\tilde{X}_1, \tilde{\omega}_1, \tilde{E}_1, \tilde{h}_1).$$

Proof. $f^*\omega_2$ is a degenerate Kähler form on X_1 and f^*h_2 is a singular Hermitian metric on f^*E_2 such that f^*h_2 is L^2 with respect to $f^*\omega_2$. Let $Z \subset X_1$ be the union of E and all four degeneracy sets. Let $U \supset Z$ be a small open neighborhood. By (11.2.1) the L^2-norms on both spaces in (11.4) are equivalent to the norms obtained by integration on $\tilde{X} - \tilde{U}$.

The only problem is that there might be some holomorphic sections of $\tilde{E}_2$ that are not holomorphic as sections of $\tilde{E}_1$. However,

$$\tilde{j} : \tilde{E}_2 \to \tilde{f}_* \tilde{E}_1$$

is an isomorphism outside a codimension two-set, hence an isomorphism. ∎

The following is the most important special case.

11.4 COROLLARY. *Let X be a compact complex manifold. The spaces of L^2 sections $H^0_{(2)}(\tilde{X}, \wedge^i \Omega^1_{\tilde{X}})$ and $H^0_{(2)}(\tilde{X}, \omega^k_{\tilde{X}})$ $(k \geq 0)$ are independent of the choice of the Kähler metric and are bimeromorphic invariants.* ∎

The following generalization of (6.4) is the main result of this section.

11.5 THEOREM. [Demailly82,92] *Let M be a line bundle on a smooth projective variety X, and $\tilde{X}$ the universal cover of X. Assume that $M \equiv N + \Delta$ where N is nef and big, and (X, Δ) is klt. Let h be a Hermitian metric on M. Then*

(11.5.1) $\mathcal{H}^{(n,i)}_{(2)}(\tilde{X}, \tilde{M}) = 0$ for $i > 0$, and

(11.5.2) $h^0(X, M \otimes K_X) = \dim_{\pi_1(X)} H^0_{(2)}(\tilde{X}, \tilde{M} \otimes K_{\tilde{X}})$.

Proof. By (10.8) $h^0(X, M \otimes K_X) = \chi(X, M \otimes K_X)$. Furthermore (cf. (6.2.2)),

$$H^0_{(2)}(\tilde{X}, \tilde{M} \otimes K_{\tilde{X}}) = \mathcal{H}^{(n,0)}_{(2)}(\tilde{X}, \tilde{M}).$$

Thus (11.5.1) implies (11.5.2).

Let H be ample on X. For $r \gg 1$ we can write $rN \equiv H + E$ where E is effective. Thus

$$(11.5.3) \quad M \equiv \frac{1}{r+s}(sN + H) + \frac{1}{r+s}E + \Delta.$$

Here $sM + H$ is ample and $E/(r+s) + \Delta$ is klt for $s \gg 1$. Choose a metric h_1 on $H(sN)$ whose curvature is everywhere positive. As in (10.6) we can use (11.5.3) to define a singular metric h_2 on M such that the corresponding curvatures satisfy the relationship

$$\Theta(h_2) = \frac{1}{r+s}\Theta(h_1) \quad \text{on the complement of } E + \Delta.$$

The estimates of [Andreotti-Vesentini65] and [Hörmander66] for the $\bar\partial$ operator have been generalized by [Demailly82] to the case of singular metrics.

11.5.4 THEOREM. [Demailly82, 5.1] *Let (Y, ω) be a complete Kähler manifold. Let M be a line bundle on Y with a Hermitian metric h and a singular Hermitian metric h_2 that is L^2 with respect to ω. Assume that its curvature form satisfies*

$$\frac{i}{2\pi}\partial\bar\partial \log h_2(f, \bar f) \geq \epsilon\omega \quad \textit{for some } \epsilon > 0.$$

(In our case the metric is C^∞ outside a subvariety $D \subset Y$ and the above condition should then hold outside D.)

Let v be an (n, i) form on Y with values in M such that $\bar\partial(v) = 0$ and

$$\int h_2(v, \bar v) < \infty.$$

Assume that $i > 0$. Then there is an $(n, i-1)$ form u on Y with values in M such that $\bar\partial(u) = v$ and

$$\int h_2(u, \bar u) < \infty. \quad \blacksquare$$

This is not exactly what we want since we measure the L^2 norm of forms in the metric h and not in h_2. Fortunately this does not cause much trouble.

The Hodge star operator (using ω and h) defines a conjugate linear isomorphism

$$*\colon \Omega_X^{(n,i)} \otimes M \cong \Omega_X^{(0,n-i)} \otimes M^{-1}.$$

We can use this to define a new complex structure on $\Omega_X^{(n,i)} \otimes M$ while keeping the metrics h and h_2.

Choose $v \in \mathcal{H}_{(2)}^{(n,i)}(\tilde{X}, \tilde{M}, \tilde{h})$. v is a harmonic (n, i) form, thus $*v$ is a harmonic $(0, n-i)$ form. Therefore the coefficients of $*v$ are antiholomorphic. In the new complex structure v becomes holomorphic, thus (11.2) implies that

$$\int \tilde{h}(v, \bar{v}) < \infty \Rightarrow \int h_2(v, \bar{v}) < \infty.$$

Thus (11.5.4) applies (to $Y = \tilde{X}$) and we obtain u as above. By construction $h_2 \geq \epsilon h$ for some $\epsilon > 0$, thus

$$\int h_2(u, \bar{u}) < \infty \Rightarrow \int \tilde{h}(u, \bar{u}) < \infty.$$

Therefore $v = \bar{\partial}(u)$ for an L^2-form u. This implies that $v = 0$ (cf. [Gromov91, 1.1.C']). ∎

CHAPTER 12

Rational Singularities and Hodge Theory

This chapter is rather technical and its results are not used elsewhere in these notes.

In the previous discussions about vanishing theorems, the starting point was the surjectivity of the natural map $H^i(X, \mathbb{C}) \to H^i(X, \mathcal{O}_X)$ for smooth projective varieties over $\mathbb{C}$. As in (10.11) it is frequently desirable to have vanishing theorems on singular spaces as well. By (9.12) the surjectivity of $H^i(X, \mathbb{C}) \to H^i(X, \mathcal{O}_X)$ for a class of singular varieties implies some vanishing theorems. We consider the case when X has rational singularities.

Let X be an arbitrary proper variety over $\mathbb{C}$. Let $\mathbb{C}_X$ denote the constant sheaf on X and $\mathcal{O}_{X^{an}}$ the holomorphic structure sheaf. We have a natural map $\mathbb{C}_X \to \mathcal{O}_{X^{an}}$ since every constant function is holomorphic. Taking Čech cohomology of both sides we obtain

$$H^i(X^{an}, \mathbb{C}_X) \to H^i(X^{an}, \mathcal{O}_{X^{an}}).$$

Of course one could take the algebraic versions as well, but the cohomology of the constant sheaf in the Zariski topology is very uninteresting. By the GAGA principle [Serre56],

$$H^i(X^{an}, \mathcal{O}_{X^{an}}) \cong H^i(X^{alg}, \mathcal{O}_{X^{alg}}).$$

I will be sloppy and write

$$H^i(X, \mathbb{C}) \to H^i(X, \mathcal{O}_X), \tag{12.1}$$

where it is understood that $H^i(X, \mathbb{C})$ is meant in the analytic topology, and for $H^i(X, \mathcal{O}_X)$ we use the analytic or the Zariski topology, whichever is more convenient.

We cannot expect (12.1) to be surjective for arbitrary singular varieties. One of the most natural restrictions is the notion of rational singularities.

12.2 Definition. Let X be a variety over a field of characteristic zero. We say that X has *rational singularities* if $R^i f_* \mathcal{O}_{X'} = 0$ for every resolution of singularities $f : X' \to X$ and for every $i > 0$.

It is sufficient to check this for one resolution of singularities. (See, e.g., [KKMS73, pp. 50–51].)

By the Leray spectral sequence, $H^i(X', \mathcal{O}_{X'}) = H^i(X, \mathcal{O}_X)$. Thus if X has rational singularities, then the coherent cohomology of the structure sheaf does not "see" the singularities. This suggests the following.

12.3 THEOREM. *Let X be a proper variety over $\mathbb{C}$. Assume that X has only rational singularities. Then*

$$H^i(X, \mathbb{C}) \to H^i(X, \mathcal{O}_X)$$

is surjective for every i.

At first sight the proof seems clear. Let $f : X' \to X$ be a resolution of singularities. $H^i(X', \mathcal{O}_{X'}) = H^i(X, \mathcal{O}_X)$ for every i, and we know that $H^i(X', \mathbb{C}) \to H^i(X', \mathcal{O}_{X'})$ is surjective. Unfortunately the natural map $H^i(X, \mathbb{C}) \to H^i(X', \mathbb{C})$ goes the wrong way. One would need some results comparing the Leray spectral sequence

$$H^i(X, R^j f_* \mathbb{C}_{X'}) \Rightarrow H^{i+j}(X', \mathbb{C})$$

and the Hodge filtration on $H^{i+j}(X', \mathbb{C})$.

The actual proof of (12.3) is rather technical and it relies on the general theory of De Rham complexes on singular spaces. In order to motivate this approach let us start with a suitably slanted view of the proof in the smooth case. The usual proof starts with the De Rham theorem representing Čech cohomology by De Rham cohomology. We want to reflect the complex structure better, thus we consider the holomorphic De Rham resolution of the constant sheaf $\mathbb{C}_X$ (where $\dim X = n$):

$$0 \to \mathbb{C}_X \to \mathcal{O}_X = \Omega_X^0 \xrightarrow{d} \Omega_X^1 \xrightarrow{d} \dots \xrightarrow{d} \Omega_X^n \longrightarrow 0.$$

This complex is exact, thus we can expect that the cohomology of $\mathbb{C}_X$ can be recovered from the "tail" of the complex

$$\Omega_X^{\cdot} = [\Omega_X^0 \xrightarrow{d} \Omega_X^1 \xrightarrow{d} \dots \xrightarrow{d} \Omega_X^n \longrightarrow 0].$$

The process of recovery is called hypercohomology (see, e.g., [Griffiths-Harris78, p. 445] for a very elementary treatment). One important point to note is that $\Omega_X^{\cdot}$ is a complex of coherent analytic sheaves, thus by the GAGA principle we should expect that its hypercohomology is the same as the hypercohomology of the corresponding algebraic De Rham complex:

$$\mathbb{H}^i(X^{an}, \Omega^{\cdot}_{X^{an}}) = \mathbb{H}^i(X^{alg}, \Omega^{\cdot}_{X^{alg}}).$$

This makes algebraists very happy since the right-hand side is defined over any field. Therefore the algebraic De Rham complex provides a replacement for $H^i(X, \mathbb{C})$ for varieties defined over an arbitrary field.

Let

$$C^{\cdot} = C^0 \to C^1 \to \cdots$$

be a complex of sheaves. C^0 can be viewed as a complex whose only nonzero term is in degree zero. C^0 is a quotient complex of $C^{\cdot}$. Applying this to $\Omega^{\cdot}_X$ we obtain a natural map

$$\text{(12.4)} \quad \mathbb{H}^i(X^{alg}, \Omega^{\cdot}_{X^{alg}}) \to H^i(X, \mathcal{O}_X),$$

which is the algebraic replacement for the map (12.1).

With these explanations in mind, the problem facing us has three parts:

(12.5.1) Find a suitable generalization $\underline{\Omega}^{\cdot}_X$ of the De Rham complex for singular varieties.

(12.5.2) Show that the corresponding map (12.4) is surjective.

(12.5.3) Show that if X has rational singularities, then $\underline{\Omega}^0_X \cong \mathcal{O}_X$.

Strictly speaking, the first step is still unsolved. Instead, one can construct a class of complexes that are "quasi isomorphic" to each other. In particular, they have the same hypercohomologies, which is all that matters for our purposes.

This seems very unsatisfactory at first, but one should remember that even in the classical case we cannot speak of *the* D Rham complex. Some people prefer the version with C^∞ coefficients, but others frequently use L^2 or distribution coefficients. All these are viewed as incarnations of *the* De Rham complex, which is a fiction of our imagination (except for people who believe that derived categories actually exist).

A further difficulty is that the corresponding map (12.4) is not surjective, unless one chooses a more complicated filtration on $\underline{\Omega}^{\cdot}_X$. This makes the "correct" $\underline{\Omega}^0_X$ into a complex, which entails only minor technical complications.

These generalized De Rham complexes were first constructed by [DuBois81]. A simplified version follows from the general theory of cubic resolutions of [GNPP88]. I list the pertinent facts, without trying to explain how the construction comes about. All the properties are either straightforward from the definitions or are explicitly mentioned in [DuBois81].

12.6 Facts. Let X be a scheme of finite type over $\mathbb{C}$. There is a complex $\underline{\Omega}_X^0$ (unique up to quasi isomorphism), called the *zeroth graded piece* of the De Rham complex, with the following properties.

(12.6.1) Let $f : Y \to X$ be a proper morphism, Y smooth. There are natural maps $\mathcal{O}_X \to \underline{\Omega}_X^0 \to R^{\cdot}f_*\mathcal{O}_Y$ such that the composite is the natural morphism $\mathcal{O}_X \to R^{\cdot}f_*\mathcal{O}_Y$.

(12.6.2) If $H \subset X$ is a general member of a base-point-free linear system, then

$$\underline{\Omega}_H^0 \cong \mathcal{O}_H \otimes \underline{\Omega}_X^0,$$

where $\cong$ means a quasi isomorphism. (The choice of H depends on the choice of the complex that represents $\underline{\Omega}_X^0$.)

(12.6.3) If X is proper then the natural map (coming from $\mathbb{C}_X \to \mathcal{O}_X \to \underline{\Omega}_X^0$)

$$H^i(X, \mathbb{C}) \to \mathbb{H}^i(X, \underline{\Omega}_X^0) \quad \text{is surjective.}$$

12.7 Definition. [Steenbrink83] We say that X has *Du Bois singularities* if $\mathcal{O}_X \to \underline{\Omega}_X^0$ is a quasi isomorphism. (By definition this means that the complex $\underline{\Omega}_X^0$ is exact, except in degree zero where the cohomology sheaf is $\mathcal{O}_X$.)

It is easy to see that smooth points are Du Bois.

If X has Du Bois singularities and X is proper, then by (12.6.3) $H^i(X, \mathbb{C}) \to H^i(X, \mathcal{O}_X)$ is surjective.

12.8 THEOREM. *Let Y be a projective and reduced scheme with Du Bois singularities and $f : Y \to X$ a morphism to a projective scheme such that the natural map*

$$\mathcal{O}_X \to R^{\cdot}f_*\mathcal{O}_Y \quad \textit{is a split injection.}$$

Then X has Du Bois singularities as well.

Proof. Let $F^i(X)$ be the ith sheaf cohomology group of $\underline{\Omega}_X^0$. Let

$$Z = \operatorname{Supp}(F^0/\mathcal{O}_X) \cup \bigcup_{i>0} \operatorname{Supp} F^i \subset X$$

be the non–Du Bois locus of X. We need to prove that Z is empty. Let $H \subset X$ be a general member of a base-point-free linear system. Then

$$\mathcal{O}_H \cong \mathcal{O}_H \otimes \mathcal{O}_X \quad \text{and} \quad R^{\cdot}f_*\mathcal{O}_{f^{-1}(H)} \cong \mathcal{O}_H \otimes R^{\cdot}f_*\mathcal{O}_Y.$$

Thus

$$\mathcal{O}_H \to R^{\cdot} f_* \mathcal{O}_{f^{-1}(H)}$$

is also a split injection. By (12.6.2),

$$F^i(H) \cong \mathcal{O}_H \otimes F^i(X) \quad \text{and} \quad F^i(f^{-1}(H)) \cong \mathcal{O}_{f^{-1}(H)} \otimes F^i(Y).$$

The second isomorphism implies that $f^{-1}(H)$ has Du Bois singularities. Thus by taking repeated hyperplane sections we may assume that $\dim Z \leq 0$.

The hypercohomology groups of $\underline{\Omega}^0_X$ are computed by a spectral sequence whose E_2 terms are $E_2^{ij} = H^i(X, F^j)$. We prove that this spectral sequence degenerates at E_2.

By assumption $\dim \operatorname{Supp} F^i \leq 0$ for $i \geq 1$ and $\dim \operatorname{Supp}(F^0/\mathcal{O}_X) \leq 0$. Thus $E_2^{ij} = 0$ if $i \geq 1$ and $j \geq 1$. Therefore the only possible nonzero differentials are

$$d^{0,j}_{j+1} : E^{0,j}_{j+1} \to E^{j+1,0}_{j+1}.$$

$\mathcal{O}_X \to R^{\cdot} f_* \mathcal{O}_Y$ is a split injection, thus the composite

$$H^i(X, \mathcal{O}_X) \xrightarrow{p_i} \mathbb{H}^i(X, \underline{\Omega}^0_X) \to \mathbb{H}^i(X, R^{\cdot} f_* \mathcal{O}_Y),$$

is an injection and so p_i is also an injection. Therefore $d^{0,j}_{j+1} = 0$ for every $j \geq 0$ and the above spectral sequence degenerates at E_2. We obtain that

$$\text{(12.8.1)} \quad \mathbb{H}^i(X, \underline{\Omega}^0_X) = H^i(X, F^0) + H^0(X, F^i) \quad (i \geq 1).$$

$H^i(X, \mathbb{C}) \to \mathbb{H}^i(X, \underline{\Omega}^0_X)$ factors through the first summand of the right hand side of (12.8.1), and hence $H^0(X, F^i) = 0$ for $i > 0$.

By assumption $\mathcal{O}_X \to F^0 \subset f_* \mathcal{O}_Y$ splits, hence $\mathcal{O}_X \to F^0$ is also a split injection which is an isomorphism at the smooth points of X. $f_* \mathcal{O}_Y$ is torsion free, so $\mathcal{O}_X = F^0$. Hence $\mathcal{O}_X \to \underline{\Omega}^0_X$ is an isomorphism on all cohomology sheaves, as was to be proved. ∎

12.8.2 Corollary. *Let $f : Y \to X$ be a finite and dominant morphism between normal and projective varieties. Assume that Y has Du Bois singularities. Then X also has Du Bois singularities.*

Proof. $R^i f_* \mathcal{O}_Y = 0$ for $i > 0$ and the normalized trace map splits $\mathcal{O}_X \to f_* \mathcal{O}_Y$ since X is normal. ∎

12.9 COROLLARY. *(Kollár-Steenbrink) Let X be a projective variety over $\mathbb{C}$ with rational singularities. Then X has Du Bois singularities.*

Proof. Let $f : Y \to X$ be a resolution of singularities. $\mathcal{O}_X \to R^{\cdot} f_* \mathcal{O}_Y$ is a quasi isomorphism by definition. Smooth points are Du Bois, thus by (12.8) X has Du Bois singularities as well. ∎

12.9.1 Remarks. (12.9.1.1) It is undoubtably true that rational singularities are Du Bois. The missing ingredient in the above proof is a compactification theorem. It is not known whether a variety with rational singularities admits a compactification with rational singularities.

(12.9.1.2) The above proof shows that if X has semirational singularities (cf. [Kollár et al.92, 12.2.1]), then X has Du Bois singularities.

The following is the surjectivity theorem implied by (12.3) using the method of (9.12). klt and lc are defined in (10.1). $L^{[k]}$ denotes the double dual of the kth tensor power and (a, b) denotes the greatest common divisor of a and b.

12.10 THEOREM. *Let X be a normal and proper variety, L a rank 1 reflexive sheaf on X, and D_i different irreducible Weil divisors on X. Assume that $L^{[m]} \cong \mathcal{O}_X(\sum d_j D_j)$ for some integers $1 \leq d_j < m$. Let $m_j = m/(m, d_j)$. Assume finally that*

$$\left(X, \sum \left(1 - \frac{1}{m_j}\right) D_j\right) \quad \textit{is klt.}$$

Then for every $i \geq 0$ and $n_j \geq 0$, the natural map

$$H^i\left(X, L^{[-1]}\left(-\sum n_j D_j\right)\right) \to H^i(X, L^{[-1]}) \quad \textit{is surjective.}$$

Proof. Let $p : Y \to X$ the normalization of the cyclic cover corresponding to $L^{[m]} \cong \mathcal{O}_X(\sum d_j D_j)$. By [Kollár et al.92, 20.2], $K_Y = p^*(K_X + \sum(1 - m_j^{-1}) D_j)$ and therefore by [ibid., 20.3] Y is klt. (In the present case all the coefficients are less than 1, so plt is the same as klt.) By [KaMaMa87, 1-3-6], Y has rational singularities. (12.9) and (9.12) imply (12.10). ∎

12.10.1 Remark. The conjecture [Kollár et al.92, 1.13] predicts that one can replace klt by lc in the statement.

Part IV

Automorphic Forms Revisited

CHAPTER 13

The Method of Gromov

One of the drawbacks of (5.22) is that it assumes the existence of bounded holomorphic functions on M. In several interesting cases this is not satisfied or not known. One such example is when X is an Abelian variety and $M = \mathbb{C}^n$. In this chapter I will present the method of [Gromov91], which does not assume anything about the existence of holomorphic functions on M. The method is fairly general and can best be formulated for group actions on topological spaces.

13.1 Notation. Let M be a topological space and L a complex line bundle on M. Let Γ be a countable discrete group acting on M and assume that the action lifts to an action on L. We do not assume that the action is free. Let $h(\, , \,)$ be a Γ-invariant continuous Hermitian metric on L and f a section of L. As before, set $\|f(x)\|^2 = h(f(x), f(x))$.

The action of Γ identifies the $\mathbb{C}$-vectorspace L_x with $L_{\gamma x}$. We denote this isomorphism by $\gamma^* : L_{\gamma x} \to L_x$. Thus we can view $\{\gamma^* f(\gamma x) | \gamma \in \Gamma\}$ as a function $\Gamma \to L_x$. Fixing a one-to-one correspondence $\Gamma \leftrightarrow \mathbb{N}$ makes the collection of all $f(\gamma x)$ into a sequence with values in L_x. If we fix an isomorphism $L_x \cong \mathbb{C}$, we obtain a sequence of complex numbers $\{a_i\}$. A different isomorphism gives the sequence $\{ca_i\}$ for some $c \in \mathbb{C} - \{0\}$.

We say that f is ℓ^p *on orbits* of Γ if the above constructed sequence of complex numbers $\{a_i\}$ is ℓ^p for every $x \in M$. This clearly does not depend on the choice of the isomorphism $L_x \cong \mathbb{C}$. It is equivalent to assuming that $\{\|f(\gamma x)\| : \gamma \in \Gamma\}$ is ℓ^p.

If $\{\|f(\gamma x)\| : \gamma \in \Gamma\}$ is ℓ^p for some $p < \infty$, then every nonzero vector $v \in L_x$ occurs only finitely many times as $\gamma^* f(\gamma x)$. Set $L_x^* = L_x - \{0\}$. We can encode the unordered sequence $\{\gamma^* f(\gamma x) | \gamma \in \Gamma\}$ as a function $F_x : L_x^* \to \mathbb{N}$, where $F_x(v)$ counts the number of occurrences of $v \in L_x^*$ among the values $\gamma^* f(\gamma x) \in L_x$. (The zeros are left out intentionally.)

If f is ℓ^p on orbits of Γ, then f is ℓ^q on orbits of Γ for every $q \geq p$. Also, if f is ℓ^p on orbits of Γ, then f^m is $\ell^{p/m}$ on orbits of Γ as a section of L^m. Thus if we are willing to go from L to a power of L we can restrict our attention to the case $p = 1$.

Assume now that f is ℓ^1 on orbits of Γ. For every $k \geq 1$ we define the *Poincaré series*

$$P(f^k)(x) = \sum_{\gamma \in \Gamma} \gamma^* f^k(\gamma x),$$

which we view this time as a section of L^k. By assumption the sum is absolutely convergent for every $x \in M$ and $k \geq 1$ and thus $P(f^k)$ is a Γ-invariant section of L^k. If M is a complex space and f is holomorphic, then $P(f^k)$ is holomorphic (5.15). In general, however, it need not be continuous.

We start with a very simple result.

13.2 THEOREM. *Notation as above. Let f be a section of L which is ℓ^1 on orbits of Γ. Assume that f is not identically zero on Γx. Then $P(f^k)(x) \neq 0$ for infinitely many k.*

Proof. This is a special case of the following lemma by setting $\{a_i\} = \{\gamma^* f(\gamma x)\}$. (We can ignore b_i.)

13.2.1 LEMMA. *Let a_i and b_i be ℓ^1-sequences of complex numbers, not all zero. Then there are infinitely many $n \in \mathbb{N}$ such that $\sum a_i^n \neq 0$ and $\sum b_i^n \neq 0$.*

Proof. After rearranging and multiplying by constants we may assume that $|a_1| = \ldots = |a_s| = 1$, $|a_i| < 1$ for $i > s$, $|b_1| = \ldots = |b_t| = 1$ and $|b_i| < 1$ for $i > t$. Let θ_i $(i = 1, \ldots, s+t)$ be the arguments of $a_1, \ldots, a_s, b_1, \ldots, b_t$. By the Dirichlet approximation theorem (see, e.g., [Hardy-Wright79, 11.12]) for every $\epsilon > 0$ there are infinitely many $n \in \mathbb{N}$ and integers r_i such that $|n\theta_i - 2\pi r_i| < \epsilon$. If in addition $n \gg 1$ then

$$\begin{aligned} &|a_i^n - 1| < \epsilon \quad \text{for } 1 \leq i \leq s, \qquad \sum_{i>s} |a_i|^n < \epsilon, \quad \text{and} \\ &|b_i^n - 1| < \epsilon \quad \text{for } 1 \leq i \leq t, \qquad \sum_{i>t} |b_i|^n < \epsilon. \end{aligned} \tag{13.2.2}$$

Thus

$$\left| s - \sum_{j=1}^{\infty} a_j^n \right| < (s+1)\epsilon \quad \text{and} \quad \left| t - \sum_{j=1}^{\infty} a_j^n \right| < (t+1)\epsilon. \quad \blacksquare$$

13.2.3 COROLLARY. *Notation as above. Assume that Γ has a compact fundamental domain and that L_M is generated by holomorphic L^2 sections.*

Then L_M^m is generated by Γ-invariant holomorphic sections for infinitely many $m > 0$. ∎

The more interesting question is how to separate different Γ-orbits using Poincaré series.

13.3 Notation. Fix a section f of L which is ℓ^1 on orbits. Let

$$R = \sum R_k \subset \sum H^0(M, L_M)^\Gamma$$

be the subalgebra generated by the Poincaré series $P(f^k)$. By definition,

$$(13.3.1) \quad R_k = \left\langle \prod_i P(f^{k_i}) \,\middle|\, \sum k_i = k \right\rangle.$$

As in (5.19) let $\mathrm{Bs}\,|R_k|$ denote the common zero set of the sections in R_k and $\phi_{R_k} : M \setminus \mathrm{Bs}\,|R_k| \to \mathbb{P}$ the map induced by $|R_k|$. (ϕ_{R_k} descends to a map on $\Gamma \backslash M$, which we also denote by ϕ_{R_k}.) Our aim is to understand the fibers of ϕ_{R_k}, at least for k sufficiently divisible.

Let $E_k(x) = \phi_{R_k}^{-1}(\phi_{R_k}(x)) \cup \mathrm{Bs}\,|R_k|$. If g_i is a basis of R_k, then $E_k(x)$ is defined by the equations $g_i(x)g_j(z) = g_i(z)g_j(x)$ $(z \in M)$. Hence $E_k(x)$ is a closed analytic subspace if the g_i are holomorphic.

The next lemma is just a reformulation of the definition.

13.4 Lemma. *Notation as above. The following are equivalent:*

(13.4.1) $\phi_{R_k}(x) = \phi_{R_k}(y)$.

(13.4.2) There is an isomorphism $\sigma_k : L_x^k \to L_y^k$ such that for every sequence $k = k_1 + \cdots + k_r$ we have

$$\sigma_k \prod_i P(f^{k_i})(x) = \prod_i P(f^{k_i})(y). \quad ∎$$

(13.4.2) can be translated into the more manageable property (13.5.1).

13.5 Proposition. *Notation as above. Let f be a section of L which is ℓ^1 on orbits of Γ and $x, y \in M$ two points. Let $F_x : L_x^* \to \mathbb{N}$ resp. $F_y : L_y^* \to \mathbb{N}$ be the corresponding unordered sequences. The following are equivalent:*

(13.5.1) There is an isomorphism $\sigma : L_x \to L_y$ such that $F_x = F_y \circ \sigma$.

(13.5.2) There is an isomorphism $\sigma : L_x \to L_y$ such that for every sequence $k_1, \ldots, k_r$

$$\sigma^{\sum k_i} \prod_i P(f^{k_i})(x) = \prod_i P(f^{k_i})(y).$$

(13.5.3) For every k there is an isomorphism $\sigma(k) : L_x^k \to L_y^k$ such that for every sequence $k = k_1 + \cdots + k_r$

$$\sigma(k) \prod_i P(f^{k_i})(x) = \prod_i P(f^{k_i})(y).$$

(13.5.4) There is an isomorphism $\sigma : L_x \to L_y$ such that $\sigma^n P(f^n)(x) = P(f^n)(y)$ for every n.

Proof. It is clear that (13.5.1)$\Rightarrow$ (13.5.2) $\Rightarrow$ (13.5.3).

(13.5.3) $\Rightarrow$ (13.5.4) can be seen as follows. Let $S \subset \mathbb{N}$ be the semigroup generated by those natural numbers s such that $P(f^s)(x) \neq 0$. If $k \in S$ then $\sigma(k)$ is uniquely determined by the assumption (13.5.3) and therefore $\sigma(k_1) \otimes \sigma(k_2) = \sigma(k_1 + k_2)$ if $k_1, k_2 \in S$. Fix an isomorphism $L_x \cong \mathbb{C}$. This makes $\sigma(k)$ into a homomorphism $\sigma(\ ) : S \to \mathbb{C}^*$, which necessarily extends to a homomorphism $\sigma(\ ) : \mathbb{Z} \to \mathbb{C}^*$. Take $\sigma = \sigma(1)$.

In order to see (13.5.4) $\Rightarrow$ (13.5.1) fix identifications $\mathbb{C} \cong L_x \cong L_y$. Thus we have a statement about two functions $F_x : \mathbb{C}^* \to \mathbb{N}$ and $F_y : \mathbb{C}^* \to \mathbb{N}$, which amounts to two unordered sequences $\{a_i\}$ and $\{b_i\}$ of complex numbers. The required implication is equivalent to the following.

13.6 Proposition. *Let $\{a_i\}$ and $\{b_i\}$ be ℓ^1-sequences of complex numbers, all different from zero. Assume that*

$$(13.6.1) \quad \sum a_i^n = \sum b_i^n \quad \textit{for all } n.$$

Then, (after suitable reordering) $a_i = b_i$ for every i.

Proof. Let $c = \max\{|a_i|, |b_i|\}$. Dividing both sequences by c we may assume that $|a_1| = \ldots = |a_s| = 1$, $|a_i| < 1$ for $i > s$, $|b_1| = \ldots = |b_t| = 1$, $|b_i| < 1$ for $i > t$ and $s + t > 0$. By choosing $n = n(\epsilon)$ suitably we may assume that the inequalities (13.2.2) are all satisfied. Thus for any fixed k we obtain that

$$\left| \sum_{i=1}^{\infty} a_i^{n+k} - \sum_{i=1}^{s} a_i^k \right| < (s+1)\epsilon \quad , \quad \textit{and}$$

$$\left| \sum_{i=1}^{\infty} b_i^{n+k} - \sum_{i=1}^{t} b_i^k \right| < (t+1)\epsilon.$$

Letting ϵ go to zero gives

$$(13.6.2) \quad \sum_{i=1}^{s} a_i^k = \sum_{i=1}^{t} b_i^k \quad \text{for every } k.$$

By adding a few zeros to one of the finite sequences in (13.6.2), we may assume that $s = t$. The elementary symmetric functions in s variables can be expressed in terms of the sums of powers (see, e.g., [v.d.Waerden71, §33, Exercise 2]). Therefore the sequences $(a_1, \dots, a_s)$ and $(b_1, \dots, b_t)$ are the roots of the same polynomial, hence they agree up to a permutation.

We can remove the first s terms of both sequences without destroying the validity of the assumptions. Continue as before to obtain the result. ∎

13.7 PROPOSITION. *Notation as in (13.1). Assume that f is ℓ^1 on orbits of Γ and it is not identically zero on any orbit. Assume furthermore that $P(f^k)$ is continuous for every $k \geq 1$.*

Fix $x \in M$. Let $E(x) \subset M$ be the set of those $y \in M$ such that there is an isomorphism $\sigma : L_x \to L_y$ such that $F_x = F_y \circ \sigma$.

Then $E(x)$ is closed and Γ-invariant in M.

Proof. Let $R = \sum R_k$ be as in (13.3). By (13.2), $\operatorname{Bs}|R| = \cap_k \operatorname{Bs}|R_k| = \emptyset$. $E_k(x) \subset M$ is closed, and by (13.4–5), $E(x) = \cap_k E_k(x)$. ∎

13.8 THEOREM. *Notation and assumptions as in (13.7). Let $E^0(x) \subset E(x)$ be the connected component of $E(x)$ containing x. Then $E^0(x) \cap \Gamma x$ is finite.*

Proof. Let $U \subset M$ be the open set where $f \neq 0$. Replacing x by γx we may assume that $x \in U$. For every $z \in U$ and $\gamma \in \Gamma$ the function $J(\gamma, z) = f(\gamma z)/f(z)$ is well defined and

$$(13.8.1) \quad J(\gamma^{-1}, \gamma z) = J(\gamma, z)^{-1}.$$

Assume that $z \in E(x)$. By (13.5) $J(\gamma, z) = f(\gamma z)/f(z)$ is an element of the countable set

$$\{0\} \cup \{f(\gamma_1 x)/f(\gamma_2 x) | \gamma_1, \gamma_2 \in \Gamma\}.$$

Thus $J(\gamma, z)$ is constant on every connected component of $U \cap E(x)$. We claim that in fact $E^0(x) \subset U$. Indeed, $E^0(x) \cap U$ is open in $E^0(x)$ and we need to show that it is also closed. Let $y \in E^0(x)$ be in the closure of $E^0(x) \cap U$. By continuity we still have that $f(\gamma y) = J(\gamma, x) f(y)$. If $y \notin U$ then $f(y) = 0$ and f vanishes on Γy, a contradiction.

Thus we obtain that $f(\gamma y) = J(\gamma, x) f(y)$ for every $y \in E^0(x)$. Let $\delta x \in E^0(x)$ for some $\delta \in \Gamma$. Then

$$J(\delta, x)^{-1} = J(\delta^{-1}, \delta x) = J(\delta^{-1}, x),$$

where the first equality follows from (13.8.1) and the second from $\delta x \in E^0(x)$. Thus

$$|f(\delta x)| + |f(\delta^{-1}x)| = \left(|J(\delta, x)| + |J(\delta^{-1}, x)|\right) |f(x)| \\ \geq 2|f(x)|.$$

Since f is ℓ^1 on Γx, this implies that $\Gamma x \cap E^0(x)$ is finite. ∎

13.8.2 COROLLARY. *Notation and assumptions as in (13.8). Assume in addition that the Γ action is discontinuous. Then $E^0(x) \to \Gamma\backslash M$ is proper.*

Proof. $E(x)$ is Γ-invariant and $E^0(x)$ is a connected component. Thus if $\gamma E^0(x) \cap E^0(x) \neq \emptyset$ then $\gamma E^0(x) = E^0(x)$. Let $\Gamma_x \subset \Gamma$ denote the stabilizer of $E^0(x)$. By (13.8) Γ_x is finite and $E^0(x) \to \Gamma\backslash M$ is the quotient morphism $E^0(x) \to \Gamma_x\backslash E^0(x)$, thus it has finite fibers.

Since Γ acts discontinuously, $p : M \to \Gamma\backslash M$ has the following property.

If $K \subset \Gamma\backslash M$ is a sufficiently small compact set, then its preimage in M is the disjoint union of compact sets.

Therefore if $F \subset M$ is closed and $p|F$ has finite fibers then $p|F$ is proper. ∎

The following is the main result of this section.

13.9 THEOREM. [Gromov91] *Let M be a normal complex space and Γ a discontinuous group of biholomorphisms. Set $X = \Gamma\backslash M$. Assume that Γ has a compact fundamental domain. Let L_M be a line bundle on M with a Γ-action and $f \in H^0(M, L_M)$ an L^p-section for some $p > 0$. Let $\phi_{R_k} : M \setminus \operatorname{Bs}|R_k| \to \mathbb{P}$ be the morphism given by weight k products of the Poincaré series $P(f^i)$: $i = 1, 2, \ldots, k$.*

(13.9.1) There is an $N = N(M, L_M, f)$ such that the fiber of $\phi_{R_{kN}}$ containing x is $E(x)$ for $k \geq 1$ and for any $x \in M \setminus \operatorname{Bs}|R_k|$. Let $\bar{E}(x)$ be the closure of $E(x)$ in M.

(13.9.2) $\bar{E}(x) \subset M$ is a closed analytic subspace and every irreducible component of $\bar{E}(x)$ is compact.

(13.9.3) Assume in addition that M is simply connected and Γ acts freely. Then $\dim \operatorname{im} \phi_{R_N} \geq \dim \operatorname{Sh}(X)$.

Proof. Let $B \subset M$ be the union of those orbits on which f vanishes. By (13.2) B is the common zero set of all the Poincaré series $P(f^k)$; thus it is a closed analytic subset. $M \setminus B$ satisfies the assumptions of (13.8). Pick $x \in M \setminus B$. Let $E(x)$ be as in (13.7) and $E_k(x)$ as in (13.3). By (13.4–5), $E(x) \cup B = \cap_k E_k(x)$; hence $E(x) \cup B$ is a closed analytic subset of M. Since X is compact, any descending chain of Γ-invariant closed analytic subsets of M stabilizes. This shows (13.9.1).

$\bar{E}(X) \subset M$ is the union of some irreducible components of the analytic subset $E(X) \cup B$, hence itself an analytic subset. Therefore the closure $\bar{E}^0(x)$ is also analytic. By (13.8.2), $\bar{E}^0(x) \to X$ has finite fibers; thus it is proper.

Since X is proper, $\bar{E}^0(x)$ is proper.

Finally let $Z \subset X$ be an irreducible component of a general fiber of ϕ_{R_N}. The preimage of Z in M is one of the subspaces $E(x)$. The irreducible components of $\bar{E}(x)$ are compact, thus $\operatorname{im}[\pi_1(\bar{Z}) \to \pi_1(X)]$ is finite, hence $\operatorname{sh}_X(Z) =$ point. This shows (13.9.3). ∎

13.9.4 Complement. Let us drop the assumption that Γ has a compact fundamental domain. Consider instead the algebra R of Poincaré series and let us look at the possible zero sets $Z(R)$ of ideals of R. If $Z(R)$ satisfies the descending chain condition, then the above proof shows that (13.9.1) still holds and $\bar{E}^0(x) \to X$ is proper.

These conditions are satisfied if X has the structure of an algebraic variety and every Poincaré series is algebraic.

13.10 COROLLARY. *Let X be a normal compact complex space with generically large fundamental group, M the universal cover of X. Let L_X be a line bundle on X and L_M the pullback of L_X to M. Assume that L_M has a nonzero L^p section for some $p > 0$. Then L_X is big, hence X is Moishezon.*

If X is smooth and $L_X = K_X$, then X is of general type.

Proof. X has generically large fundamental group, thus there is no positive dimensional proper complex subspace through a general point of M. Thus by (13.9) the general fiber of the meromorphic map $\phi_{R_N} : X \dashrightarrow \mathbb{P}$ is finite. This gives an injection $\phi_{R_N}^* \mathcal{O}_{\mathbb{P}}(1) \to L^N$, which shows that L is big.

X is Moishezon since it admits a generically finite meromorphic map onto a projective variety. If $L = K_X$ then K_X is big, which is the definition of general type. ∎

In order to get ampleness of K_X we recall (without proof) two results from higher dimensional algebraic geometry.

13.11 THEOREM. *Let X be a smooth projective variety. Then X contains a rational curve if one of the following conditions is satisfied:*

(13.11.1) [Mori82, 1.4] *There is a curve $C \subset X$ such that $C \cdot K_X < 0$.*

(13.11.2) [Kawamata91] *There is a curve $C \subset X$ and an effective divisor $E \subset X$ such that $C \cdot K_X = 0$ and $C \cdot E < 0$.* ∎

13.12 COROLLARY. *Let X be a smooth projective variety and M the universal cover of X. Assume that M does not contain positive dimensional* *compact complex subspaces (e.g., M is Stein). Assume furthermore that K_M^m has a nonzero L^p section for some $m, p > 0$. Then K_X is ample.*

Proof. By (13.10), X is of general type. X cannot contain rational curves, thus by (13.11.1) K_X is nef. The base-point-free theorem (see, e.g., [CKM88, 9.3]) shows that K_X^m is generated by global sections for $m \gg 1$. Let $f : X \to Y$ be the Stein factorization of the map given by sections of K_X^m. We need to show that it has zero dimensional fibers. Assume the contrary and let $C \subset X$ be contracted by f. In particular, $C \cdot K_X = 0$.

Let H_X be an ample divisor on X, thus $C \cdot H_X > 0$. Let H_Y be an arbitrary ample divisor on Y. For some $k \gg 1$ there is a member $D \in |kH_Y|$ which contains $f(H_X)$. Thus we can write $f^*D = H_X + E$ where E is effective:

$$C \cdot E = C \cdot f^*D - C \cdot H_X = -C \cdot H_X < 0.$$

By (13.11.2) X contains a rational curve, which is impossible. ∎

13.12.1 Complement. (13.12) holds more generally if X is a projective variety with log terminal singularities.

13.13 Comment. It is quite likely that in (13.12) it is sufficient to assume that X is a compact complex manifold. Even the following much stronger assertion should hold.

13.13.1 CONJECTURE. *Let X be a compact Moishezon manifold. If X is not projective then X contains a rational curve.*

This is known to be true if $\dim X \leq 3$.

[Campana94] realized that (13.9) can be used to get some information about $\chi(\mathcal{O}_X)$ for varieties of nongeneral type. The following applications were inspired by his work. See also [Peternell94].

13.14 PROPOSITION. *Let X be a smooth projective variety.*

(13.14.1) If K_X is numerically trivial and $\pi_1(X)$ is infinite, then $\chi(\mathcal{O}_X) = 0$.

(13.14.2) If K_X is nef but not big (i.e., X is not of general type) and X has generically large fundamental group, then $\chi(\mathcal{O}_X) = 0$.

13.14.3 Comment. The conjectures (4.16) and (4.18) imply that under the above assumptions X has an étale cover that is birational to a smooth family of Abelian varieties. In particular, $\chi(\mathcal{O}_X) = 0$. Thus (13.14) provides some evidence for these conjectures.

Proof. Let X be a smooth projective variety with universal cover M. Assume that $\chi(\mathcal{O}_X) \neq 0$. (6.3) applied to $E = K_X^{-1}$ gives a nonzero

L^2 harmonic $(0, i)$ form on M for some i. Hodge symmetry works on M (cf., e.g., [Gromov91, 1.1]) so we obtain a nonzero L^2 holomorphic i-form for some i.

Let $F_M \subset \Omega^i_M$ be the subsheaf generated by all L^2 holomorphic i-forms. F_M is $\pi_1(X)$-invariant, hence descends to a sheaf $F_X \subset \Omega^i_X$.

The idea of the proof is the following. (13.9) implies that F_X is "positive" and by (13.14.5) Ω^i_X/F_X is "nonnegative." Thus Ω^i_X is more positive than F_X. We have to make precise the appropriate notion of positivity.

Let r be the generic rank of F_X. We formulated (13.9) for line bundles, so consider

$$\wedge^r F_X \subset \wedge^r \Omega^i_X.$$

In general $\wedge^r F_X$ is not locally free. Choose $f : X' \to X$ such that

$$(13.14.4) \quad Q_{X'} := \wedge^r \Omega^i_{X'}/(\text{ saturation of } f^*(\wedge^r F_X)) \quad \text{is locally free.}$$

(This step can be avoided if one works out (13.9) for rank 1 torsion- free sheaves.) Set $E_{X'} =$ saturation of $\wedge^r F_{X'}$.

By (13.9.3) for $k \gg 1$ the global sections of $E^k_{X'}$ give a map whose image has dimension at least $\dim \mathrm{Sh}(X)$.

Let H be a very ample divisor on X and $C \subset X'$ the intersection of $(n-1)$ general members of $|f^*H|$. If $\pi_1(X)$ is infinite then sh_X is finite on C, thus $\deg(E_{X'}|C) > 0$. Define the constant $c(r, i)$ by the equation

$$\det\left(\wedge^r \Omega^i_X\right) = c(r, i) K_X.$$

By (13.14.5) $\deg(Q_{X'}|C) \geq 0$ which implies that

$$\begin{aligned} \deg(f^*K_X|C) &= \deg(K_{X'}|C) \\ &= c(r, i)^{-1}\left(\deg(E_{X'}|C) + \deg(Q_{X'}|C)\right) > 0. \end{aligned}$$

This means that K_X is not numerically trivial, proving (13.14.1).

In order to see (13.14.2) write $c_1(E_{X'}) = f^*A + F$, where A is an ample $\mathbb{Q}$-divisor on X and F is an effective $\mathbb{Q}$-divisor on X' (0.4.2). Then $c(r, i)K_X \equiv A + f_*(F + c_1(Q_X))$, and

$$c(r, i)^n K^n_X = A^n + \sum_{j=0}^{n-1} c(r, i)^{n-1-j} A^j K_X^{n-1-j} f_*(F + c_1(Q_X)).$$

Since F is effective, $A^j K_X^{n-1-j} f_* F \geq 0$ and $A^j K_X^{n-1-j} f_* c_1(Q_X) \geq 0$ by (13.14.5). Thus

$$K_X^n \geq c(r, i)^{-n} A^n > 0.$$

This shows (13.14.2). ∎

In the course of the proof we used the following result of [Miyaoka87]. A considerably simpler proof is given by Shepherd-Barron in [Kollár et al.92, 9.0.1].

13.14.5 THEOREM. [Miyaoka87] *Let X be a smooth projective variety of dimension n. Assume that X is not uniruled. Let Q be a quotient vector bundle of $(\Omega_X^1)^{\otimes s}$ for some $s \geq 1$. Let $L_1, \ldots, L_{n-1}$ be nef line bundles on X. Then*

$$c_1(Q) \cdot L_1 \cdots L_{n-1} \geq 0. \quad ∎$$

CHAPTER 14

Nonvanishing Theorems

In chapters 9–11 we proved several theorems which assert that under suitable conditions the higher cohomology groups of line bundles are zero. If a variety X and a line bundle L satisfy these assumptions, then

(14.1) $h^0(X, K_X \otimes L) = \chi(X, K_X \otimes L).$

A result of this type can be used in two ways.

(14.2.1) As a corollary we obtain that $\chi(X, K_X \otimes L) \geq 0$, which translates to an inequality between Chern classes of X and L by the Hirzebruch-Riemann-Roch formula (6.1.1). An application along this line is given in chapter 17.

(14.2.2) In some cases we are able to compute $\chi(X, K_X \otimes L)$ and thus also $h^0(X, K_X \otimes L)$. For this approach to work we need to understand the Chern classes of X. This is viable if X is in a well-defined small class of varieties, but seems hopeless if very little is known about the topology of X.

The aim of the nonvanishing theorems is to derive results about $h^0(X, K_X \otimes L)$ that are not readily visible from (14.1). By (14.2.1) this means that there are certain relations between the Chern classes of X and L. I expect that there is a large collection of results in this direction which have remained hidden so far.

In order to get an idea of the possible type of nonvanishing results, let us look at the precise form of Riemann-Roch (set $n = \dim X$).

(14.2.3) $$\chi(X, K_X \otimes L) = \frac{L^n}{n!} + \frac{L^{n-1} \cdot K_X}{2(n-1)!} + \cdots$$

If L is replaced by $L^{\otimes k}$ then the first term on the right-hand side of (14.2.3) is multiplied by k^n and all the other terms by at most k^{n-1}. Thus for large k the first term is dominant and nonvanishing follows. Informally speaking, if L is a "sufficiently large" ample line bundle, then $\chi(X, K_X \otimes L) > 0$. Unfortunately the precise meaning of "sufficiently large" depends on the Chern classes of X.

It turns out, however, that a very sharp answer can be given, at least conjecturally, which does not involve the Chern classes of X.

14.3 CONJECTURE. *(Ein-Lazarsfeld) Let X be a smooth projective variety and L a line bundle on X such that*

$$L^{\dim Z} \cdot Z > (\dim X)^{\dim Z}$$

for every subvariety $Z \subset X$. Then $\mathcal{O}(K_X + L)$ is generated by global sections, in particular $h^0(X, K_X + L) > 0$.

The precise bound is suggested by the example $X = \mathbb{P}^n, L = \mathcal{O}(n)$.

In dimension 1 this is well known. In dimension 2 this follows from the result of [Reider88]. In dimension 3 this is almost proved in [Ein-Lazarsfeld93; Fujita94]. [Kollár93a, 3.2] is a nonvanishing result of the type suggested by (14.3) with a bound essentially $(\dim X)^{2 \dim Z}$.

14.4 Fundamental group and nonvanishing. Assume that X has nontrivial fundamental group and let $f : X' \to X$ be an étale cover of degree d. Let L be ample on X. Then

$$\begin{aligned} h^0(X, \mathcal{O}(K_X + L)) &= \chi(X, \mathcal{O}(K_X + L)) \\ &= \frac{1}{d}\chi(X', \mathcal{O}(K_{X'} + f^*L)) \\ &= \frac{1}{d}h^0(X', \mathcal{O}(K_{X'} + f^*L)). \end{aligned}$$

Thus nonvanishing on X is equivalent to nonvanishing on X'. Observe that $(f^*L)^{\dim X} = d \cdot L^{\dim X}$, thus the pullback of L is "larger" than L. Assume that $d > (\dim X)^{\dim X}$. If we are lucky enough, then for any subvariety $Z \subset X$ the induced cover $Z' \to Z$ is connected and so, at least conjecturally, we obtain that $h^0(X, \mathcal{O}(K_X + L)) > 0$ as soon as L is ample. A precise form of this argument is worked out in [Kollár93b, sec. 8].

More generally one could hope that if every subvariety of X has a "large" fundamental group, then we get a stronger nonvanishing than (14.3). In order to formulate the strongest result, we need a definition.

14.4.1 Definition. Let Z be an irreducible variety and $f : Z' \to Z$ any proper birational morphism, Z' smooth. $\pi_1(Z')$ (resp. $\hat{\pi}_1(Z')$) does not depend on the choice of Z' and is denoted by $\pi_1(\operatorname{Res} Z)$ (resp. $\hat{\pi}_1(\operatorname{Res} Z)$).

Let $\bar{Z} \to Z$ be the normalization. f factors through $f' : Z' \to \bar{Z}$ and f' has connected fibers. Thus the natural map $\pi_1(\operatorname{Res} Z) \to \pi_1(\bar{Z})$ is surjective (and similarly for $\hat{\pi}_1$).

The aim of this section is to derive a nonvanishing result that is a combination and generalization of [Kollár93a, 3.2] and [Kollár93b, 8.3]. It turns out that one can build the above covering argument into the proof of [Kollár93a, 3.2] to obtain a much stronger result.

First I formulate the simplest case.

14.5 THEOREM. *Let X be a smooth and proper variety. Pick a very general point $x \in X$ (e.g., $x \in VG(X)$). Assume that if $x \in Z \subset X$ is an irreducible, positive dimensional subvariety, then $\hat{\pi}_1(\operatorname{Res} Z)$ is infinite. Let L be a big Cartier divisor on X.*

Then $h^0(X, \mathcal{O}(K_X + L)) > 0$.

14.5.1 Remark. Hopefully the result also holds if $\hat{\pi}_1$ is replaced by π_1. The proof, however, may require substantial changes.

14.6 Idea of proof. [Ein-Lazarsfeld93] Assume that we can write $L = E + M$, where $E \subset X$ is a smooth divisor and M is ample. The exact sequence

$$0 \to \mathcal{O}_X(K_X + M) \to \mathcal{O}_X(K_X + L) \to \mathcal{O}_E(K_E + M|E) \to 0$$

shows that

$$h^0(X, \mathcal{O}_X(K + L)) \geq h^0(E, \mathcal{O}_E(K_E + M|E)) \\ - h^1(X, \mathcal{O}(K_X + M)).$$

We are done by induction if $h^1(X, \mathcal{O}(K_X + M)) = 0$ and the inductive hypothesis applies to $(E, M|E)$.

Assume now that $\dim X = 2$ and $L^2 > 4$. By Riemann-Roch, $h^0(X, \mathcal{O}(kL)) > 2k^2$ for $k \gg 1$. Pick a point $x \in X$. Counting constants gives that there is a divisor $D \in |kL|$ such that $d = \operatorname{mult}_x D > 2k$. We consider two extreme cases.

14.6.1 D is a multiple of a smooth curve E. Set $M = L - E$ and the above argument applies.

14.6.2 D is smooth away from x. Let $p : X' \to X$ be the blow up of X at x with exceptional divisor $E \subset X'$. Let $D' \subset X'$ be the proper transform of D. Assume in addition that D' is smooth. Define L' by the formula $p^*(K_X + L) = K_{X'} + L'$. Then

$$L' \equiv E + \left(\frac{2}{d}D' + \left(1 - \frac{2k}{d}\right)p^*L\right).$$

$M' = L' - E$ is not ample in general, but $h^1(X', \mathcal{O}(K + L' - E)) = 0$ by (10.8). Note that when we restrict M' to E we lose L completely and only the fractional part remains. We are saved by the fact that by construction $M'|E$ is trivial.

In general, of course, one needs to perform a series of blowups and one ends up with an induction where the restriction of L to E is not ample, but rather the pullback of an ample divisor by a morphism about which we know very little. This complicates the inductive formulation considerably.

The following is the simplest inductive formulation that would work for me.

14.7 THEOREM. *Let $g : X \to S$ be a surjective morphism, X smooth and proper, and $U \subset S$ a dense open set. Let L be a nef and big $\mathbb{Q}$-Cartier $\mathbb{Q}$-divisor on S, N a Cartier divisor on X, M and Δ $\mathbb{Q}$-divisors on X. Let $s \in VG(S)$. Assume that*

(14.7.1) Supp Δ *is a normal crossing divisor and $\llcorner \Delta \lrcorner = \emptyset$;*

(14.7.2) if $s \in Z \subset S$ is an irreducible subvariety then

$$|\hat{\pi}_1(\operatorname{Res} Z)| \cdot (L^{\dim Z} \cdot Z) > \binom{\dim S + 1}{2}^{\dim Z};$$

(14.7.3) $N|g^{-1}(U)$ is linearly equivalent to an effective divisor;

(14.7.4) M is nef and either big on the general fiber of g or numerically trivial on X;

(14.7.5) $N \equiv K_X + \Delta + M + g^*L$.

Then $h^0(X, N) \neq 0$.

14.7.6 Remark. Since L is nef and big we can write it as $L = A + E$, where A is ample and E is effective. Set $d = \dim Z$. Then

$$L^d \cdot Z = A^d \cdot Z + \sum_{i=0}^{d-1} A^i L^{d-1-i} \cdot E \cdot Z.$$

If $Z \not\subset \operatorname{Supp} E$ then $L^d \cdot Z \geq A^d \cdot Z > 0$; thus in (14.7.2) we never encounter the undefined expression $\infty \cdot 0$.

Before we start the proof we should observe that the "normalized" self-intersection number $|\hat{\pi}_1(\operatorname{Res} Z)| \cdot (L^{\dim Z} \cdot Z)$ behaves very nicely in finite covers.

14.8 PROPOSITION. *Let $f : U \to V$ be a dominant, generically finite morphism between normal and proper varieties. Let L be a line bundle on V such that $L^{\dim V} > 0$. Then*

$$|\hat{\pi}_1(U)| \cdot (f^*L)^{\dim U} \geq |\hat{\pi}_1(V)| \cdot L^{\dim V},$$

and equality holds if f is étale.

14.8.1 Remark. It is easy to rewrite the statement such that the occurring terms are always finite. One way is

$$\frac{(f^*L)^{\dim U}}{L^{\dim V}} \geq [\hat{\pi}_1(V) : \operatorname{im}[\hat{\pi}_1(U) \to \hat{\pi}_1(V)]].$$

One could also use π_1 istead of $\hat{\pi}_1$.

Proof. Let $d = \deg f$. Then $(f^*L)^{\dim U} = d \cdot L^{\dim V}$. By (2.10) the image of $\hat{\pi}_1(U) \to \hat{\pi}_1(V)$ contains a subgroup of index at most d, thus $|\hat{\pi}_1(U)| \geq |\hat{\pi}_1(V)|/d$ and equality holds if f is étale. ∎

The following lemma is used to increase the self-intersection number of L.

14.9 LEMMA. *Notation and assumptions as in (14.7). Let $m : S' \to S$ be a finite étale morphism. By base change we obtain a commutative diagram*

$$\begin{array}{ccc} X' & \xrightarrow{g'} & S' \\ {\scriptstyle m_X}\downarrow & & \downarrow{\scriptstyle m} \\ X & \xrightarrow{g} & S. \end{array}$$

Then $h^0(X', m_X^*N) = \deg(m) h^0(X, N)$.

Proof. By (10.15.2)

$$\begin{aligned} h^0(X, N) &= h^0(S, g_*\mathcal{O}_X(N)) &&= \chi(S, g_*\mathcal{O}_X(N)), \quad \text{and} \\ h^0(X', m_X^*N) &= h^0(S', g'_*\mathcal{O}_{X'}(m_X^*N)) &&= \chi(S', m^*g_*\mathcal{O}_X(N)). \end{aligned}$$

m is étale, thus the Euler characteristic of a sheaf is multipled by the degree under pull back. ∎

(14.10) Proof of (14.7). We use induction on $\dim S$. If $\dim S = 0$ then $U = S$, hence we are done by (14.7.3).

If $\dim S \geq 1$, then by (14.7.2) and (14.8) we can take a finite étale cover $m : S' \to S$ such that

$$(14.7.2') \quad (m^*L)^{\dim S} > \binom{\dim S + 1}{2}^{\dim S}.$$

By (14.9) it is sufficient to prove that $h^0(X', m_X^* N) > 0$. Instead of using X' etc., we change notation and assume that (14.7.2') is also satisfied in addition to the assumptions (14.7.1–5).

By shrinking U we may assume that $g : (g^{-1}(U), \Delta) \to U$ is log smooth (i.e., g is smooth on $g^{-1}(U)$, on $\Delta_i \cap g^{-1}(U)$ for every irreducible $\Delta_i \subset \Delta$, on $\Delta_i \cap \Delta_j \cap g^{-1}(U)$ for every irreducible $\Delta_i, \Delta_j \subset \Delta$, etc.)

By easy dimension count we see that there is a $\mathbb{Q}$-divisor $B \equiv L$ on S such that $mB \in |mL|$ for some $m \gg 1$ and $\operatorname{mult}_s B > \binom{\dim S+1}{2}$.

From now on ϵ with a subscript stands for a very small positive number. Choose a log resolution $f_S : Y_S \to S$ and write

$$(14.10.1) \quad \epsilon_L f_S^* L \equiv A + \sum{}' p_i F_i, \qquad A \text{ ample}, \ 0 \leq p_i \ll 1.$$

(Here the $\sum'$ is supposed to remind us that the index set of this sum is not the same as the index set of subsequent sums without $'$.) We may assume that the first step in constructing the resolution is to blow up s. The corresponding divisor is denoted by F.

Consider $Y_S \times_S X \to X$. This is a log resolution of $(X, \Delta + g^*B)$ over $g^{-1}(U)$. By further blowups outside $g^{-1}(U)$ we can make it into a resolution

$$(14.10.2) \quad \begin{array}{ccccc} f : Y & \longrightarrow & Y_S \times_S X & \longrightarrow & X \\ \downarrow q & & \downarrow & & \\ Y_S & = & Y_S & & \end{array}$$

The unique irreducible component of $q^{-1}(F_i)$ which dominates $F_i \cap f_S^{-1}(U)$ is denoted by E_i or by $q^\circ(F_i)$. For notational simplicity we denote many other divisors on Y by E_j, we drop the $'$ from the sum notation to indicate this. There are three kinds of divisors denoted by E_i:

(i) $q^\circ(F_i)$; these are called U-divisors.

(ii) The birational transform of Δ_i where $s \in g(\Delta_i)$; these are called Δ-divisors.

(iii) All the other E_j have the property that $gf(E_j) \subset S \setminus U$. Such divisors are called negligible.

With this convention in mind let

$$(14.10.3)\qquad \begin{aligned} K_Y &\equiv f^*(K_X+\Delta)+\sum e_i E_i,\\ f^*g^*B &\equiv \sum b_i E_i;\\ \epsilon_L f^*g^*L &\equiv q^*A+\sum p_i E_i. \end{aligned}$$

The coefficient p_i for a U-divisor $E_i = q^\circ(F_i)$ is the same as the p_i in the formula (14.10.1). We can write

$$(14.10.4)\qquad \begin{aligned} f^*N \equiv K_Y &+ (1-c-\epsilon_L)f^*g^*L + f^*M + q^*A\\ &+\sum (cb_i - e_i + p_i)E_i. \end{aligned}$$

Set

$$(14.10.5)\qquad c = \min\left\{ \frac{e_i+1-p_i}{b_i} | s \in gf(E_i);\ b_i > 0\right\}.$$

If E_j is negligible then $s \notin gf(E_j)$, and if E_j is a Δ-divisor then $gf(E_j) = S$ thus $b_j = 0$. Therefore the value of c is determined by the coefficients of the U-divisors in (14.10.3).

By changing the p_i slightly we may assume that the minimum is achieved for exactly one index. Let us denote the corresponding divisor by E_0. By looking at the coefficient of $q^\circ(F)$ we conclude that $c < 2/(\dim S+1)$. Let

$$(14.10.6)\qquad \sum \llcorner cb_i - e_i + p_i \lrcorner E_i = E_0 + H'' - H',$$

where E_0, H', H'' are effective and without common irreducible components. If E_j is a Δ-divisor then $b_j = 0$ and $0 > e_j > -1$, thus $\llcorner cb_j - e_j + p_j \lrcorner = 0$.

If $cb_i - e_i + p_i < 0$ then $e_i > 0$ hence E_i is f-exceptional. If $cb_i - e_i + p_i \geq 1$ then $e_i + 1 - p_i \leq cb_i$, thus either $E_i = E_0$ or $s \not\subset gf(E_i)$. Therefore

$$(14.10.7)\qquad H' \text{ is } f\text{-exceptional and } \quad s \not\subset gf(H'').$$

Set $N' = f^*N + H' - H''$ and consider the exact sequence

$$(14.10.8)\qquad 0 \to \mathcal{O}_Y(N'-E_0) \to \mathcal{O}_Y(N') \to \mathcal{O}_{E_0}(N') \to 0.$$

By construction

$$N'|E_0 \equiv K_{E_0} + ((1-c-\epsilon_L)f^*g^*L + f^*M + q^*A)|E_0 + \sum\{cb_\iota - e_\iota + p_\iota\}E_\iota|E_0. \tag{14.10.9}$$

Set

$$\begin{aligned} &X_0 = E_0,\ S_0 = q(E_0), g_0 = q|E_0,\\ &L_0 = (1-c-\epsilon_L)f_S^*L + A|S_0,\\ &N_0 = N'|E_0,\\ &M_0 = f^*M|E_0,\\ &\Delta_0 = \sum\{cb_\iota - e_\iota + p_\iota\}E_\iota|E_0 \end{aligned} \tag{14.10.10}$$

We claim that all the conditions of (14.7.1–5) are satisfied by X_0, S_0, etc.
(14.7.1) is clear. $1-c-\epsilon_L \geq 1 - \dim S\binom{\dim S+1}{2}^{-1}$, thus

$$\begin{aligned} &|\hat{\pi}_1(\operatorname{Res} Z)| \cdot L_0^{\dim Z} \cdot Z\\ &\geq \left(1 - \frac{2}{\dim S+1}\right)^{\dim Z} |\hat{\pi}_1(\operatorname{Res} Z)| \cdot L^{\dim Z} \cdot Z\\ &> \left(\frac{\dim S - 1}{\dim S + 1}\right)^{\dim Z} \binom{\dim S+1}{2}^{\dim Z}\\ &= \binom{\dim S}{2}^{\dim Z}, \end{aligned}$$

hence (14.14.2) also holds. (14.14.4–5) hold by definition. Finally if $G_0 \subset X_0$ is the generic fiber of g_0, then $N'|G_0 = f^*N + H' - H''|G_0 = f^*N|G_0 + H'|G_0$ is effective.

Thus by induction

$$h^0(E_0, \mathcal{O}_{E_0}(N')) > 0. \tag{14.10.11}$$

Furthermore,

$$N' - E_0 \equiv K_Y + (1-c-\epsilon_L)f^*g^*L + f^*M + q^*A + \sum\{cb_\iota - e_\iota + p_\iota\}E_\iota. \tag{14.10.12}$$

If M is big on the general fiber of g then $f^*M + q^*A$ is nef and big, hence $h^i(N' - E_0) = 0$ for $i \geq 1$. In particular,

$$(14.10.13) \quad H^0(Y, \mathcal{O}_Y(N')) \to H^0(E_0, \mathcal{O}_{E_0}(N'))$$

is surjective and we are done.

If M is numerically trivial then

$$H^1(Y, \mathcal{O}(N' - E_0)) \to H^1(Y, \mathcal{O}(N'))$$

is injective by (10.13) since

$$(14.10.14) \quad (1 - c - \epsilon_L)f^*g^*L + q^*A = q^*\left((1 - c - \epsilon_L)f_S^*L + A\right)$$

is the pullback of a nef and big divisor from Y_S and $E_0 \subset \operatorname{Supp} q^*(f_S^*B))$. Thus again the morphism (14.10.13) is surjective. ∎

14.11 Proof of (14.5). L is big so we can write $L \equiv A + E$ where A is ample and E is effective (both $\mathbb{Q}$-divisors). Choose a log resolution $f : X' \to X$ of (X, E). Let $E' = f^*E$.

Apply (14.7) to $X = S = X'$, $g = id_{X'}$, $L = f^*A$, $\Delta = \{E'\}$, $M = 0$ and $N = K_{X'} + f^*L - \llcorner E' \lrcorner$. For U we can choose any open set such that $\mathcal{O}_U(N)$ is trivial. (14.7) implies that

$$h^0(X', \mathcal{O}(K_{X'} + f^*L - \llcorner E' \lrcorner) > 0.$$

By construction

$$\begin{aligned} &h^0(X', \mathcal{O}(K_{X'} + f^*L - \llcorner E' \lrcorner) \\ &\qquad \subset h^0(X', \mathcal{O}(K_{X'} + f^*L) = h^0(X', \mathcal{O}(K_X + L), \end{aligned}$$

which shows (14.5). ∎

Let $f : X \to S$ be a surjective morphism between smooth proper varieties. By the philosophy of Iitaka's program (see, e.g., [Mori87]), $f_*K^m_{X/S}$ tends to be "big" for $m > 0$. $f_*K^m_{X/S}$ is not a line bundle, thus (14.5) does not apply directly. It is, however, not difficult to obtain the following result.

14.12 THEOREM. [Kollár93b, 10.3] *Let $f : X \to S$ be a surjective morphism between smooth proper varieties with general fiber X_g. Assume* *that S has generically large algebraic fundamental group. Let D be a big Cartier divisor on S. Then*

$$h^0(X, \mathcal{O}(K_X + (m-1)K_{X/S} + f^*D)) > 0$$
$$\Leftrightarrow h^0(X_g, \mathcal{O}(mK_{X_g})) > 0. \quad \blacksquare$$

14.13 COROLLARY. [Kollár93b, 10.4] *Let $f : X \to S$ be a surjective morphism between smooth proper varieties with general fiber X_g. Assume that S is of general type and it has generically large algebraic fundamental group. Assume furthermore that $H^0(X_g, K_{X_g}^m) \neq 0$ for some $m \geq 2$. Then*

$$H^0(X, K_X^m) \geq H^0(S, K_S^{m-2}).$$

Proof. Choose $D = K_S$ in (14.12). We obtain that

$$h^0(S, \mathcal{O}(2K_S) \otimes f_*\mathcal{O}(mK_{X/S}))$$
$$= h^0(X, \mathcal{O}(K_X + (m-1)K_{X/S} + f^*K_S)) > 0.$$

This gives an injection

$$\mathcal{O}((m-2)K_S) \to f_*\mathcal{O}(mK_X). \quad \blacksquare$$

CHAPTER 15

Plurigenera in Etale Covers

Let X be a smooth projective variety and N a line bundle on X. The methods of vanishing and nonvanishing apply to N if we can write $N \equiv K + M + \Delta$, where M is nef and big and (X, Δ) is klt. Thus arises the question: can one write every line bundle in this form? If N is very negative then this is impossible, so we are primarily interested in line bundles of the form $K \otimes M$ where M is big. Even for such divisors there are easy counter examples.

The problem becomes more interesting if we are willing to change N and X. One form of the question is the following: Given X and N find a proper birational morphism $f : Y \to X$ and a line bundle N_Y on Y with a map $j : N_Y \to f^*N$ such that j induces an isomorphism $H^0(X, N) \cong H^0(Y, N_Y)$ and we can write $N_Y \equiv K_Y + M_Y + \Delta_Y$ where M_Y is nef and big and (Y, Δ_Y) is klt.

The following example shows that even this weaker requirement is not always possible to satisfy.

15.1 Example. Let E be an elliptic curve and $S = E \times E$. Pick a point $e \in E$. Let $X \to S$ be obtained by blowing up one point on $\{e\} \times E$ and denote by $C \subset X$ the birational transform of $\{e\} \times E$. C can be contracted to a point $p : X \to X'$. Let H be an ample divisor on X'.

Choose $N = p^*H^m = K_X \otimes (p^*H^m \otimes K_X^{-1})$ for some $m \gg 1$. Let $T \in \operatorname{Pic}(X)$ be numerically trivial. I claim that

$$H^1(X, T \otimes N) = \begin{cases} \mathbb{C} \text{ if } T|C \text{ is trivial, and} \\ 0 \text{ if } T|C \text{ is nontrivial.} \end{cases}$$

Indeed, $H^1(X', p_*(T \otimes N)) = 0$ for $m \gg 1$, thus

$$H^1(X, T \otimes N) = R^1 p_*(T \otimes N) = H^1(C, T|C).$$

This implies that $H^0(X, T \otimes N)$ is not locally constant at $T = \mathcal{O}_X$ as a function of T.

Assume that we can find $f : Y \to X$ and $j : N_Y \to f^*N$ as required. Vanishing applies to $f^*T \otimes N_Y$, thus

$$\begin{aligned} H^0(X, T \otimes N) &= H^0(Y, f^*T \otimes f^*N) \geq H^0(Y, f^*T \otimes N_Y) \\ &= \chi(Y, f^*T \otimes N_Y) = \chi(Y, N_Y) = H^0(Y, N_Y) \\ &= H^0(X, N). \end{aligned}$$

Therefore $H^0(X, T \otimes N)$ is locally constant at $T = \mathcal{O}_X$, a contradiction.

We restrict our attention to line bundles of the form $N = K_X^m$. More generally we could treat line bundles N such that $N \equiv a(K_X + \Delta) + L$ where $a > 1$, L is nef and big and (X, Δ) is klt (see [Kollár93b, 9]).

15.2 Proposition. *Let X be a smooth and proper variety. Assume that we can write $rK_X = F + B$, where F is nef and big and B is an effective normal crossing divisor. Assume that*

$$\llcorner \frac{m-1}{r} B \lrcorner \leq \frac{1}{d} \operatorname{Bs} |mdK_X|$$

for some d.

(15.2.1) There are $\mathbb{Q}$-divisors M and Δ where M is nef and big, Supp Δ *has normal crossings and $\llcorner \Delta \lrcorner = 0$ such that*

$$mK_X - \llcorner \frac{m-1}{r} B \lrcorner \equiv K_X + M + \Delta.$$

(15.2.2) $H^0(X, K_X^m(-\llcorner \frac{m-1}{r} B \lrcorner)) = H^0(X, K_X^m)$, and
(15.2.3) $H^i(X, K_X^m(-\llcorner \frac{m-1}{r} B \lrcorner)) = 0$ for $i > 0$.

Proof. We can write

$$\begin{aligned} mK_X - \llcorner \frac{m-1}{r} B \lrcorner &\equiv K_X + \frac{m-1}{r}(F + B) - \llcorner \frac{m-1}{r} B \lrcorner \\ &\equiv K_X + \frac{m-1}{r} F + \left\{ \frac{m-1}{r} B \right\}. \end{aligned}$$

Set $M = \frac{m-1}{r} F$ and $\Delta = \{\frac{m-1}{r} B\}$. This shows (15.2.1) and (10.8) implies (15.2.3). In order to see (15.2.2) write $B = \sum b_i B_i$, $\operatorname{Bs} |dmK_X| = \sum c_i B_i$ and let $\sum d_i B_i$ be any divisor in $|mK_X|$. We need to show that $d_i \geq \llcorner (m-1) b_i / r \lrcorner$. $\sum d d_i B_i$ is a member of $|dmK_X|$ thus $dd_i \geq c_i$. By assumption $c_i / d \geq \llcorner (m-1) b_i / r \lrcorner$, hence $d_i \geq \llcorner (m-1) b_i / mr \lrcorner$. ∎

15.3 Example. Let X be a smooth projective variety of general type. There is $r \gg 1$ such that $|rmK_X|$ gives a birational map. After suitable blowups we obtain $f : X' \to X$ such that $|rmK_{X'}| = |M| + B$ where $B = \operatorname{Bs}|rmK_{X'}|$ has normal crossings and $|M|$ is base-point free, hence nef and big.

$$\llcorner \frac{m-1}{mr} B \lrcorner \leq \frac{1}{r} \operatorname{Bs}|rmK_{X'}| = \frac{1}{r} B$$

is clear, thus (15.2) applies.

The following theorems use this construction to compare plurigenera of étale covers.

15.4 THEOREM. *Let $p : X' \to X$ be a finite étale morphism between smooth and proper varieties of general type. Then*

$$h^0(X, K_X^m) = \frac{1}{\deg p} h^0(X', K_{X'}^m) \quad \textit{for } m \geq 2.$$

Proof. Let $Y \to X' \to X$ be the Galois closure of $X' \to X$. $Y \to X$ and $Y \to X'$ are both finite, étale and Galois. Thus by going first from X to Y and then from Y to X' it is sufficient to consider the case when p is also Galois. This case is treated in (15.5). ∎

15.5 THEOREM. *Let X be a smooth and proper variety of general type. Let $p : Y \to X$ be a (possibly infinite) étale Galois cover with Galois group Γ. Then*

$$h^0(X, K_X^m) = \dim_\Gamma H^0_{(2)}(Y, K_Y^m) \quad \textit{for } m \geq 2.$$

Proof. Let $F_Y \subset K_Y^m$ be the subsheaf generated by L^2-sections. ($F_Y = 0$ is possible.) F_Y is Γ-invariant, hence there is a subsheaf $F_X \subset K_X^m$ such that $p^*F_X = F_Y$. After suitable blowups we may assume that F_X is locally free. (By (11.4) the spaces of sections in (15.5) remain the same.) By (13.2.3) F_X^r is generated by global sections for some $r > 0$. Choose $r \gg 1$. After some further blowups we may assume that $|mrK_X| = M + B$, where M is nef and big (in fact base-point-free), and $B = \operatorname{Bs}|mrK_X|$ is a normal crossing divisor. Since F_X^r is generated by global sections,

$$F_X^r \subset K_X^{mr}(-B), \quad \text{hence} \quad F_Y \subset K_Y^m(-p^* \llcorner (m-1)B/mr \lrcorner).$$

(15.2) shows that (11.5) (or (10.8)) applies to $K_X^m(-\llcorner(m-1)B/mr\lrcorner)$, hence

$$\begin{aligned}\dim_\Gamma H^0_{(2)}(Y, K_Y^m) &= \dim_\Gamma H^0_{(2)}(Y, K_Y^m(-p^*\llcorner(m-1)B/mr\lrcorner)) \\ &= h^0(X, K_X^m(-\llcorner(m-1)B/mr\lrcorner)) \\ &= h^0(X, K_X^m). \quad \blacksquare\end{aligned}$$

The results of section 14 allow us to find a pluricanonical divisor. The following result enables us to obtain a pluricanonical pencil.

15.6 THEOREM. *Let X be a smooth proper variety of general type. If $P_k(X) \geq 1$ and $P_m(X) \geq 1$ for some $k, m \geq 2$ then*

$$P_{k+m}(X) \geq \begin{cases} P_k(X) + P_m(X), & \text{if } \hat{\pi}_1(X) \neq 1, \\ \max\{P_k(X), P_m(X)\} + 1, & \text{if } \pi_1(X) \neq 1. \end{cases}$$

15.6.1 Remarks. (15.6.1.1) It is quite possible that the first inequality holds for $\pi_1(X) \neq 1$ as well.

(15.6.1.2) It is rather surprising that the fundamental group enters into the picture at all. It is not clear to me that $\pi_1(X) \neq 1$ is the relevant assumption, but the result fails for some simply connected varieties:

[Fletcher89, II.5.1] constructs a threefold of general type such that

$$P_1 = P_2 = P_3 = 0 \quad \text{and} \quad P_4 = \ldots = P_9 = 1.$$

Also, for any $r > 1$ there is a variety of general type X_r such that

$$P_1(X_r) = \ldots = P_r(X_r) = 1.$$

For the examples given in [Kollár93b, 8.6.1] $\dim X_r = 3r + 3$. It is not known what happens if we fix the dimension of X.

By (15.6) all these examples are simply connected. This, however, can be easily seen from the constructions as well.

Proof. Let $p : Y \to X$ be the étale cover whose existence is assumed. Assume first that p is finite; let $d = \deg p \geq 2$. By (15.4) and (15.6.2)

$$\begin{aligned}h^0(X, K_X^{k+m}) &= \frac{1}{d} h^0(Y, K_Y^{k+m}) \\ &\geq \tfrac{1}{d}\left(dh^0(X, K_X^k) + dh^0(X, K_X^m) - 1\right) \\ &= h^0(X, K_X^k) + h^0(X, K_X^m) - 1/d.\end{aligned}$$

$h^0(X, K_X^{k+m})$ is an integer, thus we are done.

For an infinite cover the argument is similar, except we use (15.5) and (15.6.3) instead of (15.4) and (15.6.2). ∎

15.6.2 LEMMA. *Let X be a normal and proper variety, L, M line bundles on X such that $h^0(X, L) > 0$ and $h^0(X, M) > 0$. Then $h^0(X, L \otimes M) \geq h^0(X, L) + h^0(X, M) - 1$.*

Proof. Adding divisors gives a morphism $\Sigma : |L| \times |M| \to |L+M|$. A given divisor can be written as a sum of effective divisors only in finitely many ways, thus Σ has finite fibers and therefore $\dim |L| + \dim |M| \leq \dim |L+M|$. (This proof also shows that if equality holds then the general member of $|L+M|$ is reducible.) ∎

15.6.2.1 Comment. The following nuts and bolts argument shows the reasons behind the result better, and it can be generalized to the non-compact case.

We can throw away the singular set of X without changing H^0, thus assume in addition that X is smooth. X is not proper any more but every global holomorphic function is constant.

Let D be the largest effective divisor on X such that every section of L and of M vanishes along D. We write the divisor of a section ϕ of L or M as $(\phi) = (\phi)' + D$. Pick sections f_L of L and f_M of M such that $(f_L)'$ and $(f_M)'$ have no divisors in common. Look at the maps induced by multiplication

$$f_L : H^0(X, M) \to H^0(X, L \otimes M) \quad \text{and}$$
$$f_M : H^0(X, L) \to H^0(X, L \otimes M).$$

If we prove that their images intersect in the one dimensional subspace $\langle f_L f_M \rangle$, then we are done.

Assume that $f_L g_M = f_M g_L$. Looking at divisors we obtain

$$(f_L)' + (g_M)' = (f_M)' + (g_L)'.$$

Thus $(f_L)' \leq (g_L)'$ which means that f_L divides g_L. Therefore f_L/g_L is a global holomorphic function on X, hence a constant.

A similar proof might work in the infinite dimensional case as well, but there are some technical problems. The main difficulty is that the image of multiplication by f_L (or by f_M) need not be closed, and common elements of the closures are harder to describe.

The following lemma goes around the above problem, but the conclusion is weaker.

15.6.3 LEMMA. *Let X be a proper and normal complex space and $p : Y \to X$ an étale Galois cover with infinite Galois group Γ. Let L_ι be line bundles on X. Assume that $H^0(X, L_1) \neq 0$ and $H^0_{(2)}(Y, p^*L_\iota) \neq 0$ for $i = 1, 2$. Then*

$$\dim_\Gamma H^0_{(2)}(Y, p^*L_1 \otimes p^*L_2) > \dim_\Gamma H^0_{(2)}(Y, p^*L_2).$$

Proof. Let $0 \neq s \in H^0(Y, p^*L_1)$ be Γ-invariant. Look at the multiplication map

$$m(s) : H^0_{(2)}(Y, p^*L_2) \to H^0_{(2)}(Y, p^*L_1 \otimes p^*L_2).$$

It is injective with closed range (11.2.3), thus by (6.11.1)

$$\dim_\Gamma \operatorname{im} m(s) = \dim_\Gamma H^0_{(2)}(Y, p^*L_2).$$

We are done if $m(s)$ is not surjective.

For $g \in H^0_{(2)}(Y, p^*L_1)$ let $(g)^\Gamma$ denote the Γ-invariant part of the divisor of zeros. We can view this as a divisor on X, and as such it has only finitely many irreducible components. Thus there is a section g such that $(g)^\Gamma \leq (g_1)^\Gamma$ for any $g_1 \in H^0_{(2)}(Y, p^*L_1)$. Assume that we have $f_\iota \in H^0_{(2)}(Y, p^*L_\iota)$ such that $sf_2 = gf_1$. $(g)^\Gamma \leq (f_2)^\Gamma$ by the choice of g, thus $(s)^\Gamma \leq (f_1)^\Gamma$. s is Γ-invarant, thus $(s) = (s)^\Gamma$. Therefore $(s) \leq (f_1)$ and f_1/s is a holomorphic section of $\mathcal{O}_Y$. By (11.2) it is also L^2. This implies that $f_1/s = 0$ if $|\Gamma| = \infty$. (This can be seen using the maximum principle. Alternatively, the method of (13.2) can be used to construct nonconstant sections of $\mathcal{O}_X$.) ∎

CHAPTER 16

Existence of Automorphic Forms

Let X be a smooth projective variety and $\pi : M \to X$ its universal covering space. The classical theory of automorphic forms deals with some cases where M is well known, for instance a bounded domain in $\mathbb{C}^n$. One can thus use the knowledge of M to construct sections of line bundles on X. The most frequenly studied case is for powers of the canonical bundle. The following more general situation is a natural extension of this.

16.1 Definition. Let X be a (smooth) projective variety and $\pi : M \to X$ a (usually infinite) covering space corresponding to a quotient Γ of $\pi_1(X)$. Let L_X be a line bundle on X with pullback L_M. By an L_X (resp. L_M) *valued automorphic form* on X (resp. on M) we mean a section of L_X (resp. a Γ-invariant section of L_M).

There is a one-to-one correspondence between sections of L_X and Γ-invariant sections of L_M, thus the two notions are equivalent.

It is sometimes convenient to allow the degenerate case $M = X$, when an L_X valued automorphic form is simply a section of L_X. In practice I use the expression "automorphic forms" only if the presence of the fundamental group has a significant bearing on the existence of sections.

The first result about existence of automorphic forms is a reformulation of earlier results.

16.2 THEOREM. *Let X be a smooth proper variety over $\mathbb{C}$ and L a big line bundle on X. Assume that X has generically large algebraic fundamental group. Then $h^0(X, K_X \otimes L) \geq 1$.*

16.3 THEOREM. *Let X be a smooth projective variety over $\mathbb{C}$. Assume that X has generically large algebraic fundamental group and X is of general type. Then*

(16.3.1) $h^0(X, K_X^{\otimes m}) \geq 1$ for $m \geq 2$;

(16.3.2) $h^0(X, K_X^{\otimes m}) \geq 2$ for $m \geq 4$.

Proof. (16.2) is a restatement of (14.5). (16.3.1) follows from (16.2) by setting $L = K_X^{\otimes(m-1)}$. (16.3.2) follows from (16.3.1) and (15.6). ∎

It would be very nice to have a result which says that under the above assumptions sections of $K_X^{\otimes m}$ separate points of X over an open and dense set for $m \geq m_0$, where m_0 is a reasonably small constant. (As far as I know, $m_0 = 3$ might already work.) [Kollár86a, 4.6] implies something along these lines. It does not use anything about the fundamental group, so the resulting bounds are rather large.

16.4 THEOREM. *Let X be a smooth projective variety over $\mathbb{C}$. Assume that X has generically large algebraic fundamental group and X is of general type. Then sections of $K_X^{\otimes m}$ generically separate points for $m \geq 10^{\dim X}$.*

Proof. Let $Z \subset X$ be a very general subvariety, and $Z' \to Z$ a resolution of singularities. Z' again has generically large algebraic fundamental group and it is of general type. Thus $H^0(Z', K_{Z'}^{\otimes 4}) \geq 2$. [Kollár86a, 4.6] implies the claim directly. ∎

If M is an arbitrary ample line bundle on a smooth projective variety X, then we cannot expect M^s to have sections for any fixed value of s. This fails already for curves. A section is, however, expected if M^s is much larger than the canonical bundle. The following theorem asserts a very rough version of this general principle.

16.5 THEOREM. *Let X be a smooth projective variety of dimension n and M an ample line bundle on X. Fix a positive integer $a(M)$ such that that $M^{a(M)} \otimes K_X^{-1}$ is ample. Let $p : Y \to X$ be the covering space corresponding to a quotient Γ of $\pi_1(X)$. Then for every $r \geq (n+2)^{n+6}(a(M)+n)$ the following holds:*

(16.5.1) M^r is very ample and p^*M^r is generated by its holomorphic L^2 sections, and

(16.5.2) the Poincaré map

$$P : (L^1 \cap H)(Y, p^*M^r) \to H^0(X, M^r) \quad \textit{is surjective.}$$

16.5.3 Remark. The lower bound for r is quite large. The exponential behavior comes from using some general results about very ampleness [Demailly93; Kollár93a; Siu93]. (Note added in proof. Recently announced very ampleness results of Demailly and Siu can be used to show thet (16.5) also holds for $r \geq (n+2)^2(2a(M) + (n+2)^2)$.)

Proof. The first step (which causes the large bound) is to note that M^r is very ample for $r \geq 2(n+3)!(a(M)+n)$. This follows from [Kollár93a] but we could have used [Demailly93] or [Siu93] in a similar way (with somewhat different bounds).

As in (5.12) fix a Γ-invariant Hermitian metric $\| \ \|$ on M. This induces a metric $\| \ \|^r$ on M^r. Fix a point $x \in X$ and $y \in p^{-1}(x)$. As a second step we find a $\phi \in (L^2 \cap H)(Y, p^*M^r)$ such that $\|\phi(y)\|^r = 1$ and $\phi|p^{-1}(x) - \{y\}$ has small ℓ^2 norm.

Let $I_x \subset \mathcal{O}_X$ (resp. $I_{x,Y} \subset \mathcal{O}_Y$) denote the ideal sheaf of $x \in X$ (resp. $p^{-1}(x) \subset Y$). We prove that

$$\begin{aligned} &\dim_\Gamma(L^2 \cap H)(Y, p^*M^r) \\ &\qquad = \dim_\Gamma(L^2 \cap H)(Y, I_{x,Y} \otimes p^*M^r) + 1, \end{aligned} \tag{16.5.4}$$

at least for $r \gg 1$. The strategy is to convert this into an equality on X. There should be a version of the L^2-index theorem for sheaves that does that. Instead of trying to work it out, we reduce the problem to a line bundle on another variety.

Let $\pi_X : X' \to X$ be the blowup of $x \in X$ and $\pi_Y : Y' \to Y$ the blowup of $p^{-1}(x) \subset Y$. Let $E_X \subset X'$ and $E_Y \subset Y'$ be the exceptional divisors (E_Y has several components). Clearly

$$\begin{aligned} H^0(X, I_x \otimes M^r) &= H^0(X', \pi_X^* M^r(-E_X)), \quad \text{and} \\ (L^2 \cap H)(Y, I_{x,Y} \otimes p^*M^r) &= (L^2 \cap H)(Y', \pi_Y^* p^* M^r(-E_Y)). \end{aligned} \tag{16.5.5}$$

I claim that for $r \gg 1$

$$\begin{aligned} \dim_\Gamma(L^2 \cap H)(Y, p^*M^r) &= h^0(X, M^r), \quad \text{and} \\ \dim_\Gamma(L^2 \cap H)(Y', \pi_Y^* p^* M^r(-E_Y)) &= h^0(X', \pi_X^* M^r(-E_X)), \end{aligned} \tag{16.5.6}$$

In order to see this, write

$$\begin{aligned} &(a + nb)\pi_X^* M - E_X \\ &\qquad \equiv K_{X'} + \pi_X^*(aM - K_X) + n(b\pi_X^* M - E_X). \end{aligned}$$

$aM - K_X$ is ample if $a \geq a(M)$ and $b\pi_X^* M - E_X$ is ample on X' if $(b-1)M$ is very ample. (If we embed X by $|(b-1)M|$, then $|(b-1)\pi_X^* M - E_X|$ gives the projection of X from the point x.) Thus for $a, b \gg 1$ (6.4) applies and we obtain (16.5.6). Therefore by (16.5.5–6)

$$\dim_\Gamma(L^2 \cap H)(Y, I_{x,Y} \otimes p^*M^r) = h^0(X, I_x \otimes M^r).$$

M is ample, hence M^r is globally generated for $r \gg 1$, and $h^0(X, M^r) = 1 + h^0(X, I_x \otimes M^r)$ for $r \gg 1$. Thus

$$\begin{aligned} \dim_\Gamma(L^2 \cap H)(Y, p^*M^r) = h^0(X, M^r) &= 1 + h^0(X, I_x \otimes M^r) \\ &= 1 + \dim_\Gamma(L^2 \cap H)(Y, I_{x,Y} \otimes p^*M^r). \end{aligned}$$

This shows (cf. (6.11)) that the Γ-dimension of the closure of the image of

$$\text{(16.5.7)} \quad \text{rest} : (L^2 \cap H)(Y, p^*M^r) \to M_y^{\otimes r} \otimes L^2(\Gamma) \quad \text{is } 1.$$

By (6.9.5) $L^2(\Gamma)$ has Γ-dimension 1, hence the restriction map (16.5.7) has dense image by (6.9.5). This proves (16.5.1).

By construction $\phi^2 \in (L^1 \cap H)(Y, p^*M^r)$ is such that $P(\phi^2)(x) \neq 0$. Therefore, the image of P generates M^r for $r \gg 1$. All the estimates for r have been effective so far.

An ineffective proof of (16.5.2) is simpler. Let

$$J = \sum J_r \subset \sum_{r=0}^{\infty} H^0(X, M^r)$$

denote the (vector space generated by the) image of the Poincaré map. I claim that J is an ideal. Indeed, if $\phi \in (L^1 \cap H)(Y, p^*M^r)$ and $g \in H^0(X, M^s)$, then $gP(\phi) = P(\phi p^* g)$.

By Hilbert's Nullstellensatz, either $J_r = H^0(X, M^r)$ for $r \gg 1$ or there is a point $x \in X$ such that every Poincaré series vanishes at x. We proved that the latter alternative is impossible, hence (16.5.2) holds.

The effective version follows directly from (16.6) applied to $L = M^r$ and $F = M^s$ for suitable s, r. (By (9.1) $h^i(X, M^k) = 0$ if $k \geq a(M)$ and $i > 0$.) ∎

16.5.8 COMPLEMENT. *Notation as above. Then for $r \gg 1$, holomorphic L^2 sections of p^*M^r separate points of Y.*

Proof. Pick $x_1 \neq x_2 \in X$. A computation analogous to the one given shows that for $r \gg 1$,

$$\begin{aligned} &\dim_\Gamma (L^2 \cap H)(Y, p^*M^r) \\ &\qquad - \dim_\Gamma (L^2 \cap H)(Y, I_{x_1,Y} \otimes I_{x_2,Y} \otimes p^*M^r) \\ &\quad = h^0(X, M^r) - h^0(X, I_{x_1} \otimes I_{x_2} \otimes M^r) = 2. \end{aligned}$$

The rest follows as before. ∎

16.6 PROPOSITION. [Mumford66, 14] *Let X be a projective variety, L a line bundle on X, and F a coherent sheaf on X. Let $s_i \in H^0(X, L)$ be sections that generate L. Assume that*

$$H^i(X, F \otimes L^{-i-1}) = 0 \quad \textit{for every } i \geq 1.$$

Then

$$\sum_i \operatorname{im}\left[H^0(X, F \otimes L^{-1}) \xrightarrow{s_i} H^0(X, F)\right] = H^0(X, F).$$

Proof. Look at the Koszul resolution

$$\cdots \to \bigwedge^2 \left(\sum_i L^{-1}\right) \to \sum_i L^{-1} \to \mathcal{O}_X \to 0.$$

Tensor it with F and split it up into short exact sequences

$$0 \to K_j \to F \otimes \bigwedge^j \left(\sum_i L^{-1}\right) \to K_{j-1} \to 0.$$

We need to prove that $H^1(X, K_1) = 0$. A descending induction shows that $H^i(X, K_i) = 0$ for $i > 0$. ∎

Part V

Other Applications and Speculations

CHAPTER 17

Applications to Abelian Varieties

The simplest examples of varieties with generically large algebraic fundamental group are those that admit a generically finite morphism to an Abelian variety. In this chapter we will study such varieties. This illustrates the use of the general methods, frequently in a simpler version.

The first result is a further improvement of [Kollár93b, 1.17]. Its predecessors include [Ueno75; Kawamata-Viehweg80; Kawamata81; Kollár86a]. ($b_1(X) = \dim H_1(X, \mathbb{Q})$ is the first Betti number.)

17.1 THEOREM. *Let X be a smooth proper variety over $\mathbb{C}$. The following are equivalent:*

(17.1.1) X is birational to an Abelian variety;

(17.1.2) $b_1(X) = 2\dim X$ and $h^0(X, K_X^{\otimes 3}) = 1$;

(17.1.3) $b_1(X) = 2\dim X$ and $h^0(X, K_X^{\otimes m}) = 1$ for some $m \geq 3$.

17.1.4 Remark. It is possible that 3 can be replaced by 2 in (17.1.2–3).

Proof. Let $\mathrm{alb}_X : X \to \mathrm{Alb}(X)$ be the Albanese morphism. We would like to prove that it is birational. This is done in several steps.

17.2 Step 1. First we prove that alb_X is surjective. Let $Z \subset \mathrm{Alb}(X)$ be the image, and $Z' \to Z$ a desingularization. By (14.13) $P_3(X) \geq P_1(Z')$, thus $P_1(Z') \leq 1$. The following argument of [Ueno75; Griffiths-Harris79, 4.14] implies that $Z = \mathrm{Alb}(X)$:

Pick a smooth point $0 \in Z$ and choose global coordinates on $\mathrm{Alb}(X)$ such that the tangent space of Z at 0 is given by $z_{k+1} = \dots = z_n = 0$. Near 0 the variety Z is given by the equations $z_{k+i} = f_i(z_1, \dots, z_k)$ and the f_i vanish to order 2 at 0. $z_1 \wedge \cdots \wedge z_k$ pulls pack to a global k-form on Z', and by assumption every global k-form is a constant multiple of it. Look at the forms

$$\omega_{ij} = z_1 \wedge \cdots \wedge z_{j-1} \wedge z_{k+i} \wedge z_{j+1} \wedge \cdots \wedge z_k.$$

Near 0 we can write

$$\omega_{ij} = \pm \frac{\partial f_i}{\partial z_j} z_1 \wedge \cdots \wedge z_k, \quad \text{and} \quad \frac{\partial f_i}{\partial z_j}(0) = 0.$$

By assumption ω_{ij} is a constant multiple of $z_1 \wedge \cdots \wedge z_k$, thus $\partial f_i/\partial z_j$ is identically zero. Therefore Z is a subgroup of $\mathrm{Alb}(X)$ near 0, hence it is globally a subgroup. The image of alb_X generates $\mathrm{Alb}(X)$, hence $Z = \mathrm{Alb}(X)$ and alb_X is surjective and generically finite.

With a little more care one can see that the Stein factorization of the map given by $|K_{Z'}|$ is the Iitaka fibration of Z' (cf. [Mori87, 3.10]).

17.3 Step 2. Let $X \to X^n \to \mathrm{Alb}(X)$ be the Stein factorization. If $X^n \to \mathrm{Alb}(X)$ is étale in codimension 1, then it is étale everywhere by the purity of branch loci. Thus X^n is itself an Abelian variety and we are done.

Otherwise let $D^n \subset X^n$ be a ramification divisor of $X^n \to \mathrm{Alb}(X)$ and $D \subset X$ its birational transform. $D \to \mathrm{Alb}(X)$ is generically finite, thus $P_1(D) \geq 1$. The global n-form $\mathrm{alb}_X^*(z_1 \wedge \cdots \wedge z_n)$ vanishes along the ramification locus of alb_X. Thus $D+B \in |K_X|$ for a suitable effective divisor B.

17.4 Step 3. Next we use that if X is not Abelian, then $P_m(X) \geq 2$ for some $m > 0$. This is due to [Ueno77; Kawamata-Viehweg80]. The proof is outlined in [Mori87, sec.3]. A simpler argument would be rather desirable.

17.5 Step 4. Let $f : X \dashrightarrow S$ be the Iitaka fibration [Mori87, secs.1–2]. Very general fibers of f have Kodaira dimension zero. Thus by step 3 they are birational to Abelian varieties, and $\mathrm{alb}_X(f^{-1}(s)) \subset \mathrm{Alb}(X)$ is an Abelian subvariety for very general $s \in S$. Therefore $\{\mathrm{alb}_X(f^{-1}(s)) | s \in S\}$ are translates of the same Abelian subvariety of A. We can take the corresponding quotient to obtain the following.

17.5.1 PROPOSITION. *Let X be a variety that admits a generically finite map to an Abelian variety. Then we can choose a suitable birational model X' of X and a morphism $f' : X' \to S'$ (which is the Iitaka fibration) satisfying the following properties.*

(17.5.1.1) The general fiber of f' is an Abelian variety. In particular $K_{X'}$ is trivial on the general fiber of f'. Furthermore $K_{X'} \equiv f'^*L' + B'$ where L' is big on S' and B' is effective. (L' and B' are $\mathbb{Q}$-divisors.)

(17.5.1.2) There is a generically finite morphism $q : S' \to A'$ where A' is an Abelian variety, $q(S')$ generates A', and $\dim A' \geq \kappa(X) + q(X) - \dim X$. ∎

Let $D' \subset D$ be the birational transform of D in X'. By further blowing up (both S' and X') we may also assume that $f'(D') \subset S'$ is a divisor. (This follows, for instance, from [Artin86, 5.2].)

17.6 Step 5. This is an auxiliary result.

17.6.1 LEMMA. *Let X be a smooth projective variety and $f : X \to S$ a morphism. Let H be a Cartier divisor on X. Assume that K_X and H are both numerically trivial on the general fiber on f. Let $D \subset X$ be an irreducible divisor such $P_1(D) > 0$. Assume finally that the image of X by the linear system $|mK_X + H|$ has dimension equal* $\dim S$ *for $m \gg 1$. (Since $mK_X + H$ is numerically trivial on the general fiber of f,* $\dim S$ *is the largest possible.)*

Then D is not contained in the base locus of $|mK_X + H|$ for infinitely many values of m.

Proof. Let $|mK_X+H| = |N|+aD+B$ where $aD+B$ is the fixed part. D and B are disjoint from the general fiber of f. After suitable blowups we obtain $p : X' \to X$ and we may assume that $|mK_{X'} + p^*H| = |N'| + aD' + B'$ where $|N'|$ is free and D' is the birational transform of D. We may also assume that D' is smooth. Let $f' : X' \to S'$ be given by $|N'|$. By assumption $f : X \to S$ and $f' : X' \to S'$ are birational, thus $f'(D') \neq S'$. Look at the exact sequence

$$0 \to K_{X'}(N') \to K_{X'}(D' + N') \to K_{D'}(N'|D') \to 0.$$

By (10.13) $H^1(X', K_{X'}(N')) \to H^1(X', K_{X'}(D'+N'))$ is injective. $N'|D'$ is effective, thus $h^0(X', K_{D'}(N'|D')) > 0$. Therefore $|K_{X'} + D' + N'|$ has a member E' which does not contain D'. $(a-1)D' + B' + E' \in |(m+1)K_{X'} + p^*H|$. Pushing this down to X we conclude that the multiplicity of D in the base locus of $|(m+1)K_X + H|$ is at most $a-1$. Iterating this procedure we eventually get that $D \not\subset \operatorname{Bs}|(m+j)K_X + H|$ for some $1 \leq j \leq a$. ∎

17.7 Step 6. Notation as in (17.5).

Let A be ample on S' and $H = -f'^*A$. The assumptions of (17.6) are satisfied, thus we can find a member $B \in |mK_{X'} - f'^*A|$ such that D' is not an irreducible component of B. After suitable blowups we may assume that $\operatorname{Supp} B$ has normal crossings only. Let

$$M = K_{X'}(-\llcorner B/m \lrcorner) \equiv f'^*A/m + \{B/m\}.$$

Look at the exact sequence

$$0 \to K_{X'} \otimes M \to K_{X'} \otimes M(D') \to K_{D'} \otimes M \to 0.$$

By (10.13) $H^1(X', K_{X'} \otimes M) \to H^1(X', K_{X'} \otimes M(D'))$ is injective. Also, $M|D \equiv f'^*A|D + \{B/m|D\}$ satisfies the assumptions of (14.7).

($A|f'(D')$ is big since we assumed that $f'(D') \subset S'$ is a divisor.) Therefore $h^0(D', K_{D'} \otimes M) > 0$. Thus $h^0(X', K_{X'} \otimes M(D')) \geq 2$. $K_{X'} \otimes M(D')$ is a subsheaf of $K_{X'}^3$, thus $P_3(X) = P_3(X') \geq 2$. ∎

Most of the previous proof was spent on trying to understand varieties that admit a generically finite morphism to an Abelian variety. Such varieties have been studied extensively by [Ueno75; Kawamata-Viehweg80; Kawamata81; Kollár86a]. Some of their properties are studied next.

17.8 Notation. For the rest of the chapter A denotes an Abelian variety and $f : X \to A$ a generically finite morphism from a smooth proper variety to A.

17.9 CONJECTURES. *Notation as above. Then*
(17.9.1) $P_2(X) \geq 2$ iff X is not birational to an Abelian variety;
(17.9.2) If X is of general type then $P_1(X) \geq 2$;
(17.9.3) If X is of general type then $\chi(X, K_X) > 0$.

17.9.4 Comments. By [Ueno75, 10.3; Mori87, 3.4] $\chi(X, K_X) > 0$ implies that $P_1(X \geq 2$.

By [Green-Lazarsfeld87] $\chi(X, K_X) \geq 0$, even if X is not of general type. Another proof is given in (17.12).

[Kollár86a, 4.4] gives an example of a non-Abelian compact complex surface S such that $q(S) = 2$ and $P_3(S) = 1$. Unfortunately, S is not Kähler and its Albanese is one dimensional.

17.9.5 Example. Let $A = E_1 \times E_2$ be the product of two elliptic curves, p_i the projections. Pick a degree 1 line bundle L on E_1 and a nontrivial 2-torsion line bundle T on E_2. Let $L_A = p_1^*L \otimes p_2^*T$. L_A^2 is generated by global sections. Take a general section s and let $f : X \to A$ be obtained by taking a square root of s (9.4). Then $f_*K_X = \mathcal{O}_A + L_A$. Thus $h^0(X, K_X) = 1$.

The following are some results related to (17.9).

17.10 THEOREM. *Notation as above. Then*
(17.10.1) $P_4(X) \geq \kappa(X) + 1 + q(X) - \dim X \geq \kappa(X) + 1$.
(17.10.2) If $\chi(X, K_X) > 0$ then $P_2(X) \geq q(X) + 1$.

Proof. We may assume that $q(X) = \dim A$.

First consider (17.10.1) and assume that X is of general type. Let T be a numerically trivial divisor on A and apply (14.5) to $L = K_X \pm f^*T$. We obtain that

$$|2K_X \pm f^*T| \neq \emptyset.$$

Adding divisors yields a morphism

$$|2K_X + f^*T| \times |2K_X - f^*T| \to |4K_X|.$$

Let T vary. A given divisor in $|4K_X|$ has only finitely many decompositions as a sum of integral divisors, hence $\dim |4K_X| \geq \dim A = q(X)$.

In general the proof uses the same idea, but needs more technical details. Let f', X', etc. be as in (17.5).

Let T be a numerically trivial divisor on A' and apply (14.7) to $f' : X' \to S'$ with the choices $N = 2K_{X'} \pm (q \circ f')^*T$, $\Delta = B'$, $M = 0$ and $L = L' \pm q^*T$. By our assumptions Δ is disjoint from the general fiber of f'. We obtain that

$$|2K_{X'} \pm (q \circ f')^*T| \neq \emptyset.$$

As before this implies that $\dim |4K_{X'}| \geq \dim A' \geq \kappa(X)+q(X)-\dim X$.

The proof of (17.10.2) is similar. Let T be a general numerically trivial divisor. By [Green-Lazarsfeld87]

$$h^0(X, \mathcal{O}(K_X \pm T)) = \chi(X, \mathcal{O}(K_X \pm T)) = \chi(X, \mathcal{O}(K_X)) > 0,$$

which yields that $P_2(X) \geq q(X) + 1$. ∎

The next result approaches the positivity properties of K_X from the point of view of Hilbert polynomials.

17.11 Definition. Let $f : X \to A$ be a generically finite morphism from a smooth and proper variety X to an Abelian variety A. Let L be a line bundle on A. $\chi(X, K_X \otimes f^*L^t)$ is a polynomial in t. It is denoted by $P(X, L, t)$. By (10.8) if L is ample and t is a positive integer, then

$$P(X, L, t) = H^0(X, K_X \otimes f^*L^t).$$

17.12 THEOREM. *Notation and assumptions as in (17.11). Assume that L is ample. Then every coefficient of $P(X, L, t)$ is nonnegative.*

Proof. For a natural number k let $m_k : A \to A$ be multiplication by k. Assume for simplicity that k is odd. There is a line bundle L_k on A such that $m_k^*L \cong L_k^k$ (see, e.g., [Mumford68, §6, Corollary 3]). Furthermore, L_k is very ample for $k \geq 5$ [ibid., §17]. By base change we obtain a commutative diagram (X_k may be disconnected)

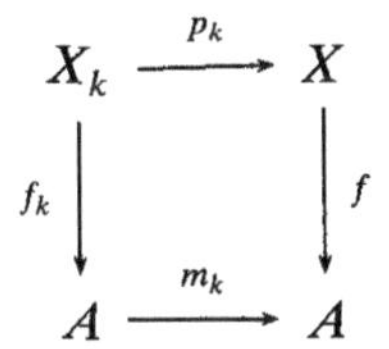

m_k and hence also p_k are étale and of degree $k^{2\dim A}$. Therefore we obtain that

$$P(X_k, L_k, kt) = k^{2\dim A} P(X, L, t).$$

As a first step we show that $\chi(X, K_X) \geq 0$. This is a result of [Green-Lazarsfeld87] and it is used in the rest of the proof:

$$\begin{aligned}\chi(X, K_X) &= P(X, L, 0)\\ &= \lim_{k\to\infty} P(X, L, 1/k) = \lim_{k\to\infty} k^{-2\dim A} P(X_k, L_k, 1)\\ &= \lim_{k\to\infty} k^{-2\dim A} H^0(X_k, K_{X_k} \otimes f_k^* L_k) \geq 0.\end{aligned}$$

A similar argument applies to the other coefficients of $P(X, L, t)$:

$$\begin{aligned}\frac{\partial^s}{\partial t^s} P(X, L, 0) &= \lim_{k\to\infty} k^s \sum_{i=0}^{s} (-1)^{s-i} \binom{s}{i} P(X, L, i/k)\\ &= \lim_{k\to\infty} k^{s-2\dim A} \sum_{i=0}^{s} (-1)^{s-i} \binom{s}{i} P(X_k, L_k, i)\\ &= \lim_{k\to\infty} k^{s-2\dim A} \sum_{i=0}^{s} (-1)^{s-i} \binom{s}{i} \chi(X_k, K_{X_k} \otimes f_k^* L_k^i).\end{aligned}$$

As we remarked, L_k is very ample for $k \geq 5$. Let $v_1, \ldots, v_s \in H^0(X_k, f_k^* L_k)$ be general sections. Let $Z_k \subset X_k$ be the intersection of their zero sets. Tensor the Koszul resolution (see, e.g., [Matsumura80, 18.D]) of $\mathcal{O}_{Z_k}$ by $K_{X_k} \otimes f_k^* L_k^s$ and take Euler characteristic to obtain that

$$\chi(Z_k, K_{Z_k}) = \sum_{i=0}^{s} (-1)^{s-i} \binom{s}{i} \chi(X_k, K_{X_k} \otimes f_k^* L_k^i).$$

Therefore

$$\frac{\partial^s}{\partial t^s} P(X, L, 0) = \lim_{k\to\infty} k^{s-2\dim A} \chi(Z_k, K_{Z_k}) \geq 0. \quad \blacksquare$$

Finally we prove a rather classical-looking result about the theta-divisors of principally polarized Abelian varieties. It is rather surprising that even in such a concrete situation the general methods say something new.

17.13 THEOREM. *Let (A, Θ) be a principally polarized Abelian variety. Then (A, Θ) is log canonical (10.1.6). In particular,*

$$\dim\{x \in A \mid \operatorname{mult}_x \Theta \geq k\} \leq \dim A - k.$$

17.13.1 Remark. The dimension estimates are sharp for every k if (A, Θ) is the product of elliptic curves. Conversely, it is possible that if equality holds for one value of k then (A, Θ) is reducible. The $k = 2$ case seems to have been around as a long-standing problem.

Proof. The presentation of the proof is due to Lazarsfeld.

The proof is one of the simplest applications of the general base-point-freeness method. Let $f : A' \to A$ be an embedded resolution of Θ. Set

$$K_{A'} = \sum e_i E_i, \quad \text{and} \quad f^*\Theta = \sum b_i E_i.$$

Let $c = \min\{(e_i + 1)/b_i\}$. We have a formal equality

$$f^*\Theta \equiv K_{A'} + (1-c)f^*\Theta + \sum(cb_i - e_i)E_i. \tag{17.13.2}$$

Let $\sum \lfloor cb_i - e_i \rfloor E_i = P - N$ where P, N are effective and without common irreducible components. If E_j is a component of N, then $e_j > 0$, thus N is exceptional. By our choice of c, P is reduced and nonempty. (17.13.2) can be rewritten as

$$f^*\Theta + N - P \equiv K_{A'} + (1-c)f^*\Theta + \sum\{cb_i - e_i\}E_i. \tag{17.13.3}$$

Set $Z = f(P)$. We obtain a commutative diagram:

$$\begin{array}{ccc} H^0(A', \mathcal{O}(f^*\Theta + N)) & \longrightarrow & H^0(P, \mathcal{O}(f^*\Theta + N)|P) \\ \cong\uparrow & & \uparrow \\ H^0(A, \mathcal{O}(\Theta)) & \xrightarrow{\ r\ } & H^0(Z, \mathcal{O}(\Theta)|Z). \end{array}$$

The left vertical arrow is an isomorphism since N is exceptional and the right vertical arrow is injective.

Assume that $c < 1$. By (10.8) we obtain that $H^1(A', \mathcal{O}(f^*\Theta + N - P)) = 0$, which implies that the top horizontal arrow is surjective. Therefore r is surjective.

On the other hand, r is zero since $Z \subset \Theta$ and the unique section of $\mathcal{O}_A(\Theta)$ vanishes along Θ. (17.13.4) leads to a contradiction. Thus $c \geq 1$. By definition this means that (A, Θ) is log canonical.

Assume that Θ has multiplicity k along a subvariety $Y \subset A$ of dimension $\dim A - k + 1$. Let us start the resolution procedure by blowing up Y, and let E_0 be the resulting exceptional divisor that dominates Y. Then $e_0 = k - 2$ and $b_0 = k$. Thus $c \leq (k - 2 + 1)/k < 1$, a contradiction. ∎

17.13.4 CLAIM. *Let $Z \subset A$ be a subscheme. Then $h^0(Z, \mathcal{O}_A(\Theta)|Z) > 0$.*

Proof. Let $\tau : A \to A$ be a general translation. Then $Z \not\subset \tau^*\Theta$, so $h^0(Z, \tau^*\mathcal{O}_A(\Theta)|Z) > 0$. The claim follows by semicontinuity. ∎

17.13.5 Remark. If C is a smooth projective curve of genus g and $(A, \Theta) = J(C)$ its Jacobian, then $u : S^{g-1} \to \Theta$ is birational and the canonical class of S^{g-1} has trivial intersection with every curve in a fiber of u. By a result of Martens, u does not contract any divisors if C is not hyperelliptic [ACGH85, IV.5.1]. This implies that

$$\Theta \text{ has } \begin{cases} \text{terminal singularities if } C \text{ is not hyperelliptic,} \\ \text{canonical singularities if } C \text{ is hyperelliptic.} \end{cases}$$

The above proof shows the following more general result. Unfortunately the only known examples to satisfy the assumptions are Abelian varieties and their quotients.

17.14 THEOREM. *Let X be a proper variety with log terminal singularities such that K_X is numerically trivial and X has large algebraic fundamental group. Let D be a nef and big Cartier divisor on X, D_g a general member of $|D|$. Then (X, D_g) is log canonical.* ∎

17.15 Problem. Let (A, Θ) be a principally polarized Abelian variety and $D \in |m\Theta|$. Is it true that $(1/m)D$ is log canonical? This would imply that

$$\dim\{x \in A \mid \operatorname{mult}_x \Theta \geq mk\} \leq \dim A - k.$$

In particular, D cannot have a singularity of multiplicity greater than $m \dim A$. One can easily see using the above argument that D cannot have an *isolated* singularity of multiplicity greater than $m \dim A$.

CHAPTER 18

Open Problems and Further Remarks

18.1 The Shafarevich conjecture. I still have not done anything about finding holomorphic functions on universal covers. The main aim of these results is rather to go around the Shafarevich conjecture.

One could say that the Shafarevich conjecture has two parts. The first part is a philosophical statement that varieties with "large" fundamental group form a rather special and important subclass of all varieties. The present notes give ample support to this assertion. The second part is a precise statement that "large" fundamental group is equivalent to the universal cover being Stein. As far as I can tell, my methods say nothing about the second part.

Conversely, the validity of the Shafarevich conjecture would imply very few of the present results. The two obvious implications are the following.

(18.1.1) In the definition of generically large fundamental group (4.6) I had to restrict myself to those subvarieties $Z \subset X$ that are not contained in a *countable* union of subvarieties D_ι. The Shafarevich conjecture would imply that a suitable *finite* union suffices.

(18.1.2) The Shafarevich conjecture implies (4.8) rather easily.

Special cases of the Shafarevich conjecture have been proved in [Gurjar87; Katzarkov94b].

18.2 The fundamental group of Sh(X). (4.8) shows that if $\pi_1(X)$ is residually finite, then after an étale cover the fundamental group of X is the same as the fundamental group of Sh(X). [Kollár93b, 4.11] shows that in general this is not true; however, in that example the kernel K of the morphism $\pi_1(X) \to \pi_1(\mathrm{Sh}(X))$ is finite and central.

[Kollár93b, 6.4.1] shows that as long as the kernel is finitely generated and residually finite we can always take an étale cover where the corresponding kernel becomes central. K is always finitely generated. If we are lucky, then K is residually finite and the above methods imply that in a suitable cover K becomes Abelian, maybe even finite. This is a very interesting question.

18.3 Algebraic versus topological fundamental group. In all of the examples I know, the notions of generically large fundamental group and generically large algebraic fundamental group are equivalent. There is, however, no good reason to believe that this is always the case.

The known constructions of varieties whose fundamental groups are not residually finite all involve large linear groups as quotients of the fundamental group [Toledo93; Catanese-Kollár92]. A very different type of construction would be needed to settle the above question.

18.4 Topological nature of the Shafarevich map. In its strongest form the problem is the following. Assume that two smooth and proper varieties X, Y are homeomorphic. Assume that $Sh(X)$ and $Sh(Y)$ both exist. Is there a homeomorphism $Sh(X) \to Sh(Y)$ such that the following diagram commutes up to homotopy?

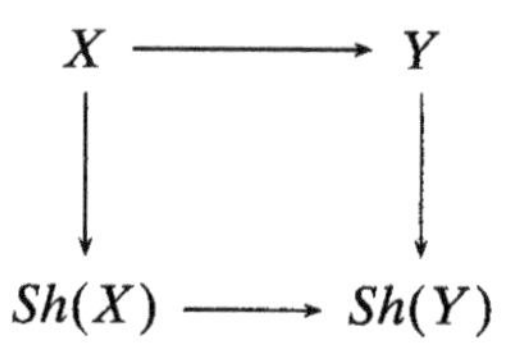

A similar question could be asked about $\mathrm{Sh}(X)$ and $\mathrm{Sh}(Y)$; however, the formulation requires a little extra care since birationally equivalent varieties are ususally not homotopy equivalent.

Some results of [Mok92] point in this direction.

A weaker version of this question asks if a deformation of X induces a deformation of $\mathrm{Sh}(X)$ in a natural way. This should be much easier.

Let $F \subset X$ be a general fiber of sh_X. Typically $h^1(F, \mathcal{O}_F) = 0$, which implies that F lifts to any small deformation X_t of X, giving a general fiber of sh_{X_t}. If $h^1(F, \mathcal{O}_F) > 0$, then $X \to \mathrm{Sh}(X)$ factors through a (not everywhere smooth) family of Abelian varieties over $\mathrm{Sh}(X)$. Again there is hope that these can be studied in detail.

18.5 Varieties with $\kappa = 0$. Let X be a smooth projective variety such that $\kappa(X) = 0$ (i.e., $h^0(X, K_X^m) \leq 1$ for every $m > 0$ and equality holds for some m). Is it true that $\pi_1(X)$ contains a finite index Abelian subgroup? This question was discussed a little in (4.16).

[Mok92] shows that under the above assumptions $\pi_1(X)$ does not have any Zariski dense and discrete representation in a semisimple algebraic group.

18.6 Universal covers of Abelian schemes. Let $f : X \to Y$ be an Abelian scheme and $\tilde{Y} \to Y$ the universal cover. Then $X \times_Y \tilde{Y} \to \tilde{Y}$ is a family of Abelian varieties with a global topological trivialization.

This gives a morphism $\tilde{Y} \to H$, where H is the Siegel upper half space parametrizing Abelian varieties with a given polarization and fixed first homology. Assume that the Shafarevich morphism $\tilde{Y} \to Sh(\tilde{Y})$ exists. We have a factorization

$$\tilde{Y} \to Sh(\tilde{Y}) \to H.$$

This shows that $\tilde{X}$ is the pullback of the universal bundle $\mathrm{Univ}_H \to H$. Furthermore, we obtain that

$$Sh(\tilde{X}) \to Sh(\tilde{Y})$$

is a vector bundle. The universal cover of a variety of general type is measure hyperbolic [Kobayashi70, IX.1]. We arrive at the following.

18.6.1 CONJECTURE. *Let X be a smooth projective variety. Then $Sh(\tilde{X})$ is a vector bundle over a measure hyperbolic Stein space.*

18.7 Kähler hyperbolic manifolds. [Gromov91] defined a Kähler manifold (M, ω) to be Kähler hyperbolic if the pullback of ω to the universal cover is d of a bounded 1-form. It is easy to see that Kähler hyperbolic implies that M has large fundamental group. The converse fails since by [ibid. 0.2.A'] the fundamental group of a Kähler hyperbolic manifold cannot be Abelian. Thus, for example, a sufficiently general, smooth, ample hypersurface X in an Abelian variety of dimension at least 3 is not Kähler hyperbolic. X has large fundamental group and ample canonical class.

From the birational point of view, Kähler hyperbolicity is not natural since it depends on the birational model chosen. It would be desirable to have a birational variant developed. The natural choice seems to be to require Gromov's condition not for a Kähler form but for a degenerate Kähler form (11.1).

18.8 Characterization of universal covers of algebraic varieties.

18.8.1 CONJECTURE. *Let X be a smooth proper variety of dimension n. Assume that X is of general type and that X has generically large fundamental group. Then $\tilde{X}$ carries a nonzero, L^2, holomorphic n-form.*

This would imply that there is a generically nondegenerate Bergman metric on $\tilde{X}$.

[Gromov91] proved that (18.8.1) is true if X is Kähler hyperbolic. In general the following weaker version holds.

18.8.2 THEOREM. *Let M be a complex manifold and Γ a group of biholomorphisms acting discretely and properly discontinuously on M such that $\Gamma\backslash M$ is compact. Assume that a sufficiently general point of M is not contained in any irreducible positive dimensional compact subspace of M. The following are equivalent:*

(18.8.2.1) There is a nonzero, L^2, holomorphic pluricanonical form on M.

(18.8.2.2) $\Gamma\backslash M$ is a proper algebraic space of general type with generically large fundamental group.

Furthermore, if Γ is residually finite then the above conditions are equivalent to the following:

(18.8.2.3) There is a nonzero, L^2, holomorphic 2-canonical form on M.

Proof. Clearly (18.8.2.3) $\Rightarrow$ (18.8.2.1). Assume (18.8.2.1). By (3.10) $\Gamma\backslash M$ is of general type. It is easy to see (cf. [Kollár93b, 2.12.2]) that $\Gamma\backslash M$ has generically large fundamental group. This proves (18.8.2.2).

Finally, assume (18.8.2.2). Set $X = \Gamma\backslash M$. By assumption $h^0(X, \mathcal{O}(mK_X)) > 0$ for some $m > 0$, and we can choose $m = 2$ if Γ is residually finite (14.5). As in (15.3) there is a birationally equivalent model X' of X and divisors L', E' with the following properties:

(18.8.2.4.1) E' is an effective integral divisor.

(18.8.2.4.2) There are q-divisors N' and Δ' such that $M' \equiv N' + \Delta'$ where N' is nef and big, $\llcorner \Delta' \lrcorner = 0$ and $\operatorname{Supp} \Delta'$ has normal crossings only.

(18.8.2.4.3) $mK_{X'} = K_{X'} + L' + E'$.

(18.8.2.4.4) $h^0(X, \mathcal{O}(mK_X)) = h^0(X', \mathcal{O}(K_{X'} + L'))$.

Take any Kähler metric $\omega_{X'}$ on X' and any Hermitian metric h' on $\mathcal{O}(L')$. From (11.5.2) we conclude that there is a nonzero $\phi' \in H^0_{(2)}(M', \mathcal{O}(K_{M'} + \tilde{L}'))$. By (11.4) this implies that $H^0_{(2)}(M, \mathcal{O}(mK_M))$ is nonzero. ∎

18.9 Nonvanishing theorems. The proof of the nonvanishing theorem (14.5) uses very heavily that X has generically large algebraic fundamental group. I see, however, no reason why the result should fail if X has generically large fundamental group. The proof will probably require an analytic approach to nonvanishing. The case when X is Kähler hyperbolic was settled by [Katzarkov93].

In another direction, one can ask the following generalization of (14.5).

18.9.1 CONJECTURE. *Let X be a smooth and proper variety, $X^0 \subset X$ and $f^0 : X^0 \to Z^0$ as in (1.8). Let $F \subset X^0$ be a general fiber of f^0. Let L be a big Cartier divisor on X such that $H^0(F, \mathcal{O}_F(K_F + L|F)) > 0$.*

Then $h^0(X, \mathcal{O}(K_X + L)) > 0$.

18.10 Kodaira dimension of varieties with generically large fundamental group. Let X be a smooth proper variety with generically large fundamental group. Is it true that $\kappa(X) \geq 0$?

This follows from general conjectures about the minimal model program. It is, however, possible that in our special case there is a direct proof.

The first step might be to prove that $c_1(K_X)$ is in the closure of the cone of effective divisors. One proof might proceed as follows.

Under the assumptions of (14.5) one should prove that $K_X + M$ is big. Then by induction $mK_X + M = K_X + ((m-1)K_X + M)$ is big. Hence K_X is the limit of the effective classes $\frac{1}{m}(mK_X + M)$.

18.11 Pluricanonical maps. Let X be a smooth proper variety of general type and with generically large algebraic fundamental group. By (16.4) the sections of $K_X^{\otimes m}$ define a birational map for $m \geq 10^{\dim X}$. As far as I know the same might be true for $m \geq 3$. From the point of view of the proof, it is more reasonable to expect a bound that is linear in $\dim X$. The question of a significantly better bound is open even for varieties generically finite over Abelian varieties.

18.12 Positivity properties. By Hirzebruch-Riemann-Roch

$$\chi(X, L \otimes K_X) = \int_X \operatorname{ch}(L) \cdot \operatorname{ch}(K_X) \cdot \operatorname{td}(X).$$

We can view this as a polynomial in L whose coefficients are the homogeneous components of $\operatorname{ch}(K_X) \cdot \operatorname{td}(X)$. The Kodaira vanishing theorem implies that $\chi(X, L \otimes K_X) \geq 0$ if L is ample, which should imply some positivity statements about $\operatorname{ch}(K_X) \cdot \operatorname{td}(X)$. For an arbitrary variety these are very little understood. If, however, X has generically large fundamental group, some easy-to-state positivity results might hold. The simplest is the following conjecture about the constant term.

18.12.1 CONJECTURE. *Let X be a smooth projective variety. Assume that X has generically large fundamental group. Then $\chi(X, \omega_X) > 0$.*

If X is Kähler hyperbolic, this was proved in [Gromov91]. If X is generically finite over an Abelian variety, $\chi(X, \omega_X) \geq 0$ by [Green-Lazarsfeld87].

More generally one can ask the following.

18.12.2 QUESTION. *Let X be a smooth projective variety. Assume that the universal cover of X is Stein. Is it true that every homogeneous component of* $\operatorname{ch}(K_X) \cdot \operatorname{td}(X)$ *is "semipositive" in some sense? (For instance,* *semipositive may mean that it is the limit of effective cycles, or that it has nonnegative intersection with ample divisors.)*

(17.12) shows something like this for varieties generically finite over an Abelian variety.

18.13 Characterization of Abelian varieties. Let X be a smooth and proper variety. Is it true that X is birational to an Abelian variety iff $H^1(X, \mathbb{C}) = \mathbb{C}^{2\dim X}$ and $H^0(X, K_X^2) = \mathbb{C}$?

(17.1) proves this with $H^0(X, K_X^3) = \mathbb{C}$ replacing $H^0(X, K_X^2) = \mathbb{C}$. (17.9.5) shows that $H^0(X, K_X^2) = \mathbb{C}$ cannot be replaced by $H^0(X, K_X) = \mathbb{C}$.

The proof of (17.1) has two parts. The first step is to show that alb_X is generically finite and the second step is to study varieties that are generically finite over an Abelian variety. At the moment, neither of these steps works with $H^0(X, K_X^2) = \mathbb{C}$.

18.14 Shafarevich maps for Kähler varieties. This was discovered independently by [Campana94]. The definitions are the same as in (3.5, 4.6). This seems a little surprising since a Kähler variety may not have any proper subvarieties, in which case the definition simply says that X has generically large fundamental group iff $\pi_1(X)$ is infinite.

Nonetheless it seems that this definition works, especially in the above degenerate case. For instance [Campana94] proves the following.

18.14.1 THEOREM. [Campana94, 2.5] *Let X be compact Kähler manifold. Assume that there is a point $x \in X$ which is not contained in any nontrivial compact subvariety. Assume furthermore that $\chi(X, \mathcal{O}_X) \neq 0$.*

Then $|\pi_1(X)| \leq 2^{\dim X - 1}$. ∎

18.15 Dimension bound for $\mathrm{sh}(X)$. Is there a numerical function $c(\Gamma)$ defined on groups such that if $\pi_1(X) \cong \Gamma$ then $\dim \mathrm{Sh}(X) \leq c(\Gamma)$? More generally, if there is a homomorphism $\pi_1(X) \to \Gamma$ with kernel H, then one could hope that $\dim \mathrm{Sh}^H(X) \leq c(\Gamma)$.

If Γ is finite then $c(\Gamma) = 0$. If Γ is Abelian then one can take $c(\Gamma) = \mathrm{rank}\,\Gamma/2$. The results of [Simpson91; Mok92; Zuo94; Jost-Zuo93; Katzarkov94a] give such bounds for certain linear groups.

18.16 Kähler class in the group cohomology. The following very interesting and deep question is due to Carlson and Toledo.

Let X be a smooth projective variety with generically large fundamental group Γ. Is there a finite index subgroup $\Gamma' < \Gamma$ and a cohomology class $\delta \in H^2(\Gamma')$ such that its image under the natural map $H^2(\Gamma') \to H^2(X, \mathbb{Q})$ is the Chern class of a nef and big divisor?

As a very special case, is it true that $H^2(\Gamma') \neq 0$ for some finite index subgroup $\Gamma' < \Gamma$?

Assume that the stronger variant holds. Let L_X be the corresponding line bundle and $L_{\tilde{X}}$ the pullback of L_X to the universal cover of X. $L_{\tilde{X}}$ is topologically trivial, and maybe even holomorphically trivial. Since L_X is big, for $m \gg 1$ we can write $L_X^m \cong K_X \otimes H(E)$, where H is ample and E is effective. This way we obtain holomorphic sections of $L_{\tilde{X}}^m$, and hence holomorphic functions on $\tilde{X}$.

18.17 The norm of the Poincaré map. Let X be a complex space with universal cover $\tilde{X} \to X$. Let M_X be a vector bundle on X, $M_{\tilde{X}}$ its pullback. Fix a Hermitian metric h on M_X. By pullback we obtain a metric $\tilde{h}$ on $M_{\tilde{X}}$. The Poincaré map is a linear functional

$$P = P(X, M) : (L^1 \cap H)(\tilde{X}, M_{\tilde{X}}) \to (L^1 \cap H)(X, M_X)$$

between two Banach spaces.

The beautiful theorem of [McMullen89,92] shows that the norm of $P(X, M)$ is less than one if $\pi_1(X)$ is not amenable and one if $\pi_1(X)$ is amenable. The proof of [McMullen92] is very clear and elementary.

18.18 Holomorphic convexity of $\tilde{X}$ with respect to line bundles. A complex space M is called "holomorphically convex" if for every sequence $\{x_i\} \subset M$ without limit points there is a holomorphic function f on M which is unbounded on $\{x_i\}$. By the Remmert reduction theorem (see, e.g., [Grauert-Remmert84, p. 221]), the Shafarevich conjecture holds for X iff $\tilde{X}$ is holomorphically convex.

Let L_X be a line bundle on X with a Hermitian metric $\| \ \|$. We say that $p : \tilde{X} \to X$ is holomorphically convex with respect to $(L, \| \ \|)$ if for every sequence $x_i \in \tilde{X}$ without limit points there is a holomorphic section f of p^*L such that the sequence $\|f(x_i)\|$ is unbounded.

[Napier90] proves that if X is projective and L is ample, then $\tilde{X}$ is holomorphically convex with respect to $(L^r, \| \ \|^r)$ for $r \gg 1$.

It is possible to derive this result from (16.5.1).

Pick any $x \in \tilde{X}$ and let f_x be an L^2-section of p^*L^r such that $\|f_x(x)\| > 1$. Since f_x is L^2, for every $\epsilon > 0$ there is a compact subset $K_x(\epsilon) \subset \tilde{X}$ such that $\|f_x(y)\| < \epsilon$ for every $y \notin K_x(\epsilon)$. Let $D \subset \tilde{X}$ be a compact fundamental domain. There is a compact set $K(D, \epsilon) \subset \tilde{X}$ and finitely many sections $f_i = f_{x_i}$ such that

$$\max\{\|f_i(y)\|\} > 1 \quad \text{if } y \in D, \text{ and}$$
$$\max\{\|f_i(y)\|\} < \epsilon \quad \text{if } y \notin K(D, \epsilon).$$

Let $T \subset \tilde{X}$ be any compact subset and

$$H(T, \epsilon) = \cup\{\gamma D: T \cap \gamma K(D, \epsilon) \neq \emptyset\}.$$

Assume that $y \notin H(T, \epsilon)$. Choose $\gamma \in \pi_1(X)$ such that $\gamma^{-1} y \in D$. Then $T \cap \gamma K(D, \epsilon) = \emptyset$, thus there is an $i = i(y)$ such that

$$\|\gamma f_i(y)\| > 1 \quad \text{and} \quad \|\gamma f_i(z)\| < \epsilon \quad \text{for } z \in T.$$

Let $x_i \in \tilde{X}$ be an unbounded sequence and $T_1 \subset T_2 \subset \cdots$ a compact exhaustion of $\tilde{X}$. By the above considerations for every i there is an $m(i)$ and a section f_i of M^r such that

$$\|f_i(x)\| < 2^{-i} \quad \text{if } x \in T_i, \text{ and}$$

$$\|f_i(x_{m(i)})\| > 2^i + \sum_{j<i} \|f_j(x_{m(i)})\|.$$

Thus $f = \sum f_i$ is a holomorphic section of M^r which is unbounded on the sequence $\{x_i\}$.

18.19 Slow growth functions on covering spaces. [Lárusson93] shows that if X is a smooth projective variety with universal covering $\tilde{X}$ and $H \subset X$ a smooth ample divisor, then every "slowly growing" holomorphic function on $\tilde{X} \times_X H$ extends to a "slowly growing" holomorphic function on $\tilde{X}$.

It is possible that this result can be used to reduce the Shafarevich conjecture to surfaces.

18.20 Nilpotent groups. [Campana93] constructed an example of a smooth projective variety whose fundamental group is nilpotent but not Abelian by finite. He also proves that if $\pi_1(X)$ is nilpotent then the Stein factorization of the Albanese morphism is the Shafarevich map of X.

18.21 A conjecture of Nori. The Shafarevich conjecture has the following interesting consequence, first investigated by [Nori83].

18.21.1 CONJECTURE. *Let X be a normal and proper variety and $Z \subset X$ a connected subvariety. Let $Z_i \to Z$ be the irreducible components of the normalization. Assume that* $\operatorname{im}[\pi_1(Z_i) \to \pi_1(X)]$ *is finite for every i. Then* $\operatorname{im}[\pi_1(Z) \to \pi_1(X)]$ *is finite.*

Recent results of [Lasell-Ramachandran94] give further evidence to this conjecture.

18.22. [Lyons-Sullivan84; Nadel90a; Napier-Ramachandran94] contain further results about universal covers of algebraic varieties that rely on methods very different from the ones used in this book.

REFERENCES

[ACGH85] E. Arbarello, M. Cornalba, P. A. Griffiths, and J. Harris, Geometry of Algebraic Curves, vol. 1. Springer, 1985.

[Ahlfors64] L. Ahlfors, *Finitely generated Kleinian groups*, Amer. J. Math. 86 (1964), 413–429.

[Andreotti-Vesentini65] A. Andreotti and E. Vesentini, *Carleman estimates for the Laplace-Beltrami equation in complex manifolds*, Publ. Math. IHES 25 (1965), 81–130.

[Arapura86] D. Arapura, *A note on Kollár's theorem*, Duke Math. J. 53 (1986), 1125–1130.

[Arapura94] D. Arapura, *Fundamental group of smooth projective varieties* (to appear), in Current Directions in Algebraic Geometry. 1992–93 MSRI Special Year in Algebraic Geometry, 1994.

[Artin69] M. Artin, *Algebraisation of formal moduli I.*, in Global Analysis, ed. D. C. Spenser and S. Iyanaga, pp. 21–72. Univ. Tokyo Press and Princeton Univ. Press, 1969.

[Artin86] M. Artin, *Néron Models*, in Arithmetic Geometry, ed. G. Cornell and J. Silverman, pp. 213–230. Springer, 1986.

[Arveson76] W. Arveson, An Invitation to C^* Algebras, GTM, vol. 39. Springer, 1976.

[Atiyah76] M. Atiyah, *Elliptic operators, discrete groups and von Neumann algebras*, Astérisque 32–33, (1976), 43–72.

[Barlet75] D. Barlet, *Espace analytique réduit des cycles analytiques complexes de dimension finie*, in Séminaire Norguet, pp. 1–158. Springer Lecture Notes, vol. 482, 1975.

[Beauville83] A. Beauville, *Some remarks on Kähler manifolds with $c_1 = 0$*, in Classification of Algebraic and Analytic Manifolds, pp. 1–26. Birkhäuser, 1983.

[Bell66] D. Bell, *Poincaré series representations of automorphic forms*, Ph. D. diss., Brown Univ., 1966.

[Bers65] L. Bers, *Automorphic forms and Poincaré series for infinitely generated Fuchsian groups*, Amer. J. Math. 87 (1965), 196–214.

[Blanchard56] M. A. Blanchard, *Sur les variétés analytiques complexes*, Ann. Sci. Ec. Norm. Sup. 73 (1956), 157–202.

[Borel63] A. Borel, *Compact Clifford-Klein forms of symmetric spaces*, Topology 2 (1963), 111–122.

[BPV84] W. Barth, C. Peters, and A. Van de Ven, Compact Complex Surfaces. Springer, 1984.

[Campana81] F. Campana, *Coréductions algébrique d'un espace analytique faiblements kählerien compact*, Inv. Math. 63 (1981), 187–223.

[Campana91] F. Campana, *On twistor spaces of the class C*, J. Diff. Geom. 33 (1991), 541–549.

[Campana93] F. Campana, *Remarks on nilpotent Kähler groups*, C. R. Acad. Sci. Paris 317 (1997), 777–780.

[Campana94] F. Campana, *Remarques sur le revêtement universel des varietes kähleriennes compactes*, Bull. Soc. Math. France 122 (1994), 255–284.

[Carlson-Toledo89] J. Carlson and D. Toledo, *Harmonic mappings of Kähler manifolds to locally symmetric spaces*, Publ. Math. IHES 69 (1989), 173–201.

[Catanese-Kollár92] F. Catanese and J. Kollár, *Trento examples 2*, in Classification of Irregular Varieties, pp. 136–139. Springer Lecture Notes, vol. 1515, 1992.

[CGM82] J. Cheeger, M. Goresky, and R. MacPherson, *L^2 cohomology and intersection homology of singular algebraic varieties*, in Seminar on Differential Geometry, ed. S.-T. Yau, pp. 303–340. Princeton Univ. Press, 1982.

[Cheeger-Gromov85] J. Cheeger and M. Gromov, *On the characteristic numbers of complete manifolds of bounded curvature and finite volume*, in Differential Geometry and Complex Analysis, ed. I. Chavel and H. Farkas, pp. 115–155. Springer, 1985.

[CKM88] H. Clemens, J. Kollár, and S. Mori, Higher Dimensional Complex Geometry, vol. 166, Astérisque, 1988.

[Conway90] J. B. Conway, A Course in Functional Analysis. Springer, 1990.

[Danilov78] V. I. Danilov, *The geometry of toric varieties*, Russian Math. Surveys 33 (1978), 97–154.

[Deligne79] P. Deligne, *Le groupe fondamental du complément d'une courbe plane n'ayant que des points doubles ordinaires est abélien*, Sém. Bourbaki, no. 543, 1979.

[Deligne-Mostow86] P. Deligne and G. D. Mostow, *Monodromy of hypergeometric functions and non-lattice integral monodromy*, Publ. Math. IHES 63 (1986), 5–90.

[Demailly82] J.-P. Demailly, *Estimations L^2 pour l'opérateur $\bar\partial \ldots$*, Ann. Sci. E.N.S. 15 (1982), 457–511.

[Demailly89] J. P. Demailly, *Une généralisation de théorème d'annulation de Kawamata-Viehweg*, C. R. Acad. Sci. Paris 309 (1989), 123–126.

[Demailly92] J.-P. Demailly, *Singular Hermitian metrics on positive line bundles*, in Complex Algebraic Varieties, pp. 87–104. Springer Lecture Notes, vol. 1507, 1992.

[Demailly93] J. P. Demailly, *A numerical criterion for very ample line bundles*, J. Diff. Geom. 37 (1993), 323–374.

[Dixmier81] J. Dixmier, Von Neumann Algebras. North-Holland, 1981.

[DuBois81] Ph. Du Bois, *Complex de De Rham filtré d'une variété singulière*, Bull. Soc. Math. France 109 (1981), 41–81.

[Earle69] C. J. Earle, *Some remarks on Poincaré series*, Comp. Math. 21 (1969), 167–176.

[Ein-Lazarsfeld93] L. Ein and R. Lazarsfeld, *Global generation of pluricanonical and adjoint linear systems on smooth projective threefolds*, Jour. AMS 6 (1993), 875–903.

[Esnault-Viehweg86] H. Esnault and E. Viehweg, *Logarithmic de Rham complexes and vanishing theorems*, Inv. Math. 86 (1986), 161–194.

[Esnault-Viehweg87] H. Esnault and E. Viehweg, *Revêtements cycliques II*, in Géométrie Algèbrique et Applications II, La Rábida, pp. 81–94. Herman, Paris, 1987.

[Esnault-Viehweg92] H. Esnault and E. Viehweg, *Lectures on vanishing theorems*, in DMV Sem., vol. 20. Birkhäuser, 1992.

[Fletcher89] A. Fletcher, *Working with weighted complete intersections* (MPI preprint) (1989).

[Fujita94] T. Fujita, *Remarks on Ein-Lazarsfeld criterion of spannedness of adjoint bundles of polarized threefold* (preprint) (1994).

[Fulton-Lazarsfeld81] W. Fulton and R. Lazarsfeld, *Connectivity and its applications in algebraic geometry*, in Algebraic Geometry, pp. 26–92. Springer Lecture Notes, vol. 862, 1981.

[GNPP88] F. Guillén, V. Navarro-Aznar, P. Pascual-Gainza, and F. Puerta, *Hyperrésolutions cubiques et descente cohomologique*. Springer Lecture Notes, vol. 1335, 1988.

[Goresky-MacPherson83] M. Goresky and R. MacPherson, *Intersection homology II*, Inv. Math. 71 (1983), 77–129.

[Goresky-MacPherson88] M. Goresky and R. MacPherson, Stratified Morse Theory. Springer, 1988.

[Grauert-Remmert84] H. Grauert and R. Remmert, Coherent Analytic Sheaves. Springer, 1984.

[Grauert-Riemenschneider70] H. Grauert and O. Riemenschneider, *Verschwindungssätze für analytische Kohomologiegruppen auf komplexen Räumen*, Invent. Math. 11 (1970), 263–292.

[Green-Lazarsfeld87] M. Green and R. Lazarsfeld, *Deformation theory, generic vanishing theorems and some conjectures of Enriques, Catanese and Beauville*, Inv. Math. 90 (1987), 389–407.

[Griffiths-Harris78] P. Griffiths and J. Harris, Principles of Algebraic Geometry. Wiley, 1978.

[Griffiths-Harris79] P. Griffiths and J. Harris, *Algebraic geometry and local differential geometry*, Ann. Sci. E.N.S. 12 (1979), 355–452.

[Gromov91] M. Gromov, *Kähler hyperbolicity and L_2-Hodge theory*, J. Diff. Geom. 33 (1991), 263–292.

[Grothendieck62] A. Grothendieck, Fondéments de la Géométrie Algébrique, Sec. Math. Paris, 1962.

[Gunning76] R. Gunning, Riemann Surfaces and Generalized Theta Functions. Springer, 1976.

[Gunning-Rossi65] R. Gunning and H. Rossi, Analytic Functions of Several Complex Variables. Prentice Hall, 1965.

[Gurjar87] R. V. Gurjar, *Coverings of algebraic varieties*, in Algebraic Geometry, Sendai, vol. 10, pp. 179–184, ed. T. Oda. Adv. Stud. Pure Math. Kinokuniya–North-Holland, 1987.

[Hardy-Wright79] G. Hardy and E. Wright, An Introduction to the Theory of Numbers, 5th ed. Clarendon, Oxford, 1979.

[Hartshorne77] R. Hartshorne, Algebraic Geometry. Springer, 1977.

[Helgason78] S. Helgason, Differential Geometry, Lie Groups and Symmetric Spaces. Academic Press, 1978.

[Hirzebruch58] F. Hirzebruch, *Automorphe Formen und der Satz von Riemann-Roch*, pp. 129–144. Symp. de Top. Alg., Univ. Nac. Mexico, 1958.

[Hirzebruch66] F. Hirzebruch, Topological Methods in Algebraic Geometry. Springer, 1966.

[Hirzebruch83] F. Hirzebruch, *Arrangements of lines and algebraic surfaces*, in Arithmetic and Geometry II, pp. 113–140. Progress in Math. 36. Birkhäuser, 1983.

[Hodge-Pedoe52] W. Hodge and D. Pedoe, Methods of Algebraic Geometry. Cambridge Univ. Press, 1952.

[Hörmander66] L. Hörmander, An Introduction to Complex Analysis in Several Variables. North-Holland, 1966.

[Iitaka71] S. Iitaka, *On D-dimension of algebraic varieties*, J. Math. Soc. Japan 23 (1971), 356–373.

[Ishida88] M.-N. Ishida, *An elliptic surface covered by Mumford's fake projective plane*, Tohoku Math. J. 40 (1988), 367–396.

[Jost-Zuo93] J. Jost and K. Zuo, *Harmonic maps into Tits buildings, factorizations of nonrigid and nonarithmetic representations of π_1 of algebraic varieties* (preprint) (1993).

[Katzarkov93] L. Katzarkov, *Nonvanishing theorems for Kähler hyperbolic varieties* (preprint) (1993).

[Katzarkov94a] L. Katzarkov, *Factorization theorems for the representations of the fundamental groups of quasiprojective varieties and some applications* (preprint) (1994).

[Katzarkov94b] L. Katzarkov, *On the Shafarevich conjecture* (preprint) (1994).

[KaMaMa87] Y. Kawamata, K. Matsuda, and K. Matsuki, *Introduction to the minimal model problem*, in Algebraic Geometry, Sendai, vol. 10, pp. 283–360, ed. T. Oda. Adv. Stud. Pure Math. Kinokuniya–North-Holland, 1987.

[Kawamata81] Y. Kawamata, *Characterisation of Abelian varieties*, Comp. Math. 43 (1981), 253–276.

[Kawamata82] Y. Kawamata, *A generalisation of Kodaira-Ramanujam's vanishing theorem*, Math. Ann. 261 (1982), 43–46.

[Kawamata91] Y. Kawamata, *On the length of an extremal rational curve*, Inv. Math. 105 (1991), 609–611.

[Kawamata-Viehweg80] Y. Kawamata and E. Viehweg, *On a characterisation of Abelian varieties in the classification theory of algebraic varieties*, Comp. Math. 41 (1980), 355–360.

[KKMS73] G. Kempf, F. Knudsen, D. Mumford, and B. Saint-Donat, *Toroidal embeddings I*. Springer Lecture Notes, vol. 339, 1973.

[Klingen90] H. Klingen, Introductory Lectures on Siegel Modular Forms. Cambridge Univ. Press, 1990.

[Kobayashi70] S. Kobayashi, Hyperbolic Manifolds and Holomorphic Mappings. M. Dekker, 1970.

[Kodaira53] K. Kodaira, *On a differential geometric method in the theory of analytic stacks*, Proc. Nat. Acad. Sci. USA 39 (1953), 1268–1273.

[Kollár86a] J. Kollár, *Higher direct images of dualizing sheaves I*, Ann. of Math. 123 (1986), 11–42.

[Kollár86b] J. Kollár, *Higher direct images of dualizing sheaves II*, Ann. of Math. 124 (1986), 171–202.

[Kollár87a] J. Kollár, *The structure of algebraic threefolds—an introduction to Mori's program*, Bull. AMS 17 (1987), 211–273.

[Kollár87b] J. Kollár, *Vanishing theorems for cohomology groups*, in Algebraic Geometry Bowdoin 1985, pp. 233–243. Proc. Symp. Pure Math., vol. 46, 1987.

[Kollár93a] J. Kollár, *Effective base point freeness*, Math. Ann. 296 (1993), 595–605.

[Kollár93b] J. Kollár, *Shafarevich maps and plurigenera of algebraic varieties*, Inv. Math. 113 (1993), 177–215.

[Kollár et al.92] J. Kollár et al., Flips and Abundance for Algebraic Threefolds, vol. 211, Astérisque, 1992.

[KoMiMo92] J. Kollár, Y. Miyaoka, and S. Mori, *Rationally connected varieties*, J. Alg. Geom. 1 (1992), 429–448.

[Lárusson93] F. Lárusson, *An extension theorem for holomorphic functions of slow growth*... (preprint) (1993).

[Lasell-Ramachandran94] B. Lasell and M. Ramachandran, *Local systems on Kähler manifolds and harmonic maps* (preprint) (1994).

[Lefschetz24] S. Lefschetz, L'Analysis Situs et la Géometrie Algébrique. Gauthier-Villars, 1924.

[Lyons-Sullivan84] T. Lyons and D. Sullivan, *Function theory, random paths and covering spaces*, J. Diff. Geom. 19 (1984), 244–323.

[Margulis91] G. A. Margulis, Discrete Subgroups of Semisimple Lie Groups. Springer, 1991.

[Matsumura80] H. Matsumura, Commutative Algebra, 2d ed. Benjamin-Cummings, 1980.

[McMullen89] C. McMullen, *Amenability, Poincaré series and quasiconformal maps*, Inv. Math. 97 (1989), 95–127.

[McMullen92] C. McMullen, *Amenable coverings of complex manifolds and holomorphic probability measures*, Inv. Math. 110 (1992), 29–37.

[Miyaoka80] Y. Miyaoka, *On the Mumford-Ramanujam vanishing theorem on a surface*, in Géométrie Algébrique, Angers, pp. 239–247. Sijthoff and Nordhoff, 1980.

[Miyaoka87] Y. Miyaoka, *The Chern classes and Kodaira dimension of a minimal variety*, in Algebraic Geometry, Sendai, vol. 10, pp. 449–476, ed. T. Oka. Adv. Stud. Pure Math. Kinokuniya–North-Holland, 1987.

[Miyaoka88] Y. Miyaoka, *On the Kodaira dimension of minimal threefolds*, Math. Ann. 281 (1988), 325–332.

[Mok92] N. Mok, *Factorizations of semi simple discrete representations of Kähler groups*, Inv. Math. 110 (1992), 557–614.

[Mori82] S. Mori, *Threefolds whose canonical bundles are not numerically effective*, Ann. of Math. 116 (1982), 133–176.

[Mori87] S. Mori, *Classification of higher-dimensional varieties*, in Algebraic Geometry, Bowdoin 1985, pp. 269–332. Proc. Symp. Pure Math., vol. 46 (1987).

[Mori-Mukai83] S. Mori and S. Mukai, *The uniruledness of the moduli space of curves of genus 11*, in Algebraic Geometry, Tokyo/Kyoto 82, pp. 334–353. Springer Lecture Notes, vol. 1016 1983.

[Mumford66] D. Mumford, Lectures on Curves on an Algebraic Surface. Princeton Univ. Press, 1966.

[Mumford68] D. Mumford, Abelian Varieties. Tata Inst. Lecture Notes, 1968.

[Mumford76] D. Mumford, Algebraic Geometry I. Springer, 1976.

[Mumford79] D. Mumford, *An algebraic surface with K ample, $K^2 = 9$, $p_g = q = 0$*, Amer. J. Math. 101 (1979), 233–244.

[Mumford-Fogarty82] D. Mumford and J. Fogarty, Geometric Invariant Theory, 2d ed. Springer, 1982.

[Nadel90a] A. M. Nadel, *Semisimplicity of the group of biholomorphisms of the universal cover of a compact complex manifold with ample canonical bundle*, Ann. Math. 132 (1990), 193–211.

[Nadel90b] A. M. Nadel, *Multiplier ideal sheaves and Kähler-Einstein metrics of positive scalar curvature*, Ann. Math. 132 (1990), 549–596.

[Nakamura75] I. Nakamura, *Complex parallelisable manifolds and their small deformations*, J. Diff. Geom., 10 (1975), 85–112.

[Namikawa93] Y. Namikawa, *On deformations of Calabi-Yau threefolds with terminal singularities* (1993). Topology (to appear).

[Namikawa94] Y. Namikawa, *Global smoothing of Calabi-Yau threefolds by flat deformations* (preprint) (1994).

[Napier90] T. Napier, *Convexity properties of coverings of smooth projective varieties*, Math. Ann. 286 (1990), 433–479.

[Napier-Ramachandran94] T. Napier and M. Ramachandran, *Structure theorems for complete Kähler manifolds and applications to Lefschetz theorems* (preprint) (1994).

[Nori 83] M. Nori, *Zariski's conjecture and related results*, Ann. Sci. ENS 16 (1983), 305–344.

[Peternell94] T. Peternell, *Minimal varieties with trivial canonical classes, I* (1994). Math. Zeitschrift (to appear).

[Poincaré1883] H. Poincaré, *Mémoires sur les fonctions Fuchsiennes*, Acta. Math. 1 (1883), 193–294.

[Ramanujam72] C. P. Ramanujam, *Remarks on the Kodaira vanishing theorem*, J. Indian Math. Soc. 36 (1972), 41–51.

[Ran94] Z. Ran, *Hodge theory and deformations of maps*, Comp. Math. (to appear) (1994).

[Reid80] M. Reid, *Canonical Threefolds*, in Géométrie Algébrique Angers, ed. A. Beauville, pp. 273–310. Sijthoff & Noordhoff, 1980.

[Reider88] I. Reider, *Vector bundles of rank 2 and linear systems on algebraic surfaces*, Ann. Math. 127 (1988), 309–316.

[Resnikoff69] H. Resnikoff, *Supplement to "Some remarks on Poincaré series,"* Comp. Math. 21 (1969), 177–181.

[Rudin66] W. Rudin, Real and Complex Analysis. McGraw-Hill, 1966.

[Rudin80] W. Rudin, Function Theory in the Unit Ball in $\mathbb{C}^n$. Springer, 1980.

[Saito90] M. Saito, *Decomposition theorem for proper Kähler morphisms*, Tohoku Math. J. 42 (1990), 127–148.

[Saito91] M. Saito, *On Kollár's conjecture*, in Several Complex Variables and Complex Geometry, pp. 509–517. Proc. Symp. Pure Math., vol. 52, 1991.

[Segal83] D. Segal, Polycyclic Groups. Cambridge Univ. Press, 1983.

[Serre56] J.-P. Serre, *Géometrie algébrique et géometrie analytique*, Ann. Inst. Fourier (1956), 1–42.

[SGA1] A. Grothendieck et al., *Revêtements étales et groupes fondamental*. Springer Lecture Notes, vol. 224, 1971.

[Shafarevich72] R. I. Shafarevich, Basic Algebraic Geometry (in Russian). Nauka, 1972.

[Siegel73] C. L. Siegel, Topics in Complex Function Theory, vol. 3. Wiley, 1973.

[Simpson91] C. Simpson, *The ubiquity of variations of Hodge structures*, in Complex Geometry and Lie Theory, Proc. Symp. Pure Math., vol. 53, 1991, pp. 329–348.

[Siu80] Y.-T. Siu, *Complex analiticity of harmonic maps and strong rigidity of complex Kähler manifolds*, Ann. of Math. 112 (1980), 73–111.

[Siu87] Y.-T. Siu, *Rigidity for Kähler manifolds and the construction of bounded holomorphic functions*, in Discrete Groups in Geometry and Analysis, pp. 124–151. Progress in Math., vol. 67. Birkhäuser, 1987.

[Siu93] Y.-T. Siu, *An effective Matsusaka big theorem* (preprint) (1993).

[Steenbrink77] J. H. M. Steenbrink, *Mixed Hodge structures on the vanishing cohomology*, in Real and Complex Singularities, pp. 525–563. Sijthoff and Noordhoff, 1976.

[Steenbrink83] J. H. M. Steenbrink, *Mixed Hodge structures associated with isolated singularities*, in Singularities, Part 2, Proc. Symp. Pure Math., vol. 40 (1983), 513–536.

[Stein56] K. Stein, *Überlagerungen holomorph-vollständiger komplexe Räume*, Arch. Math. 7 (1956), 354–361.

[Tankeev71] S. G. Tankeev, *On n-dimensional canonically polarised varieties*, Izv. A.N. SSSR 35 (1971), 31–44.

[Toledo90] D. Toledo, *Examples of fundamental groups of compact Kähler manifolds*, Bull. London Math. Soc. 22 (1990), 339–343.

[Toledo93] D. Toledo, *Projective varieties with non-residually finite fundamental group*, Publ. Math. IHES 77 (1993), 103–119.

[Ueno75] K. Ueno, *Classification theory of algebraic varieties and compact complex spaces*. Springer Lecture Notes, vol. 439, 1975.

[Ueno77] K. Ueno, Classification of Algebraic Varieties, II, pp. 693–708. Intl. Symp. Alg. Geom. Kyoto, Kinokuniya, 1977.

[v.d.Waerden71] B. L. van der Waerden, Algebra, 8th ed. Springer, 1971.

[Viehweg82] E. Viehweg, *Vanishing theorems*, J. f. r. u. a. Math. 335 (1982), 1–8.

[Wells73] R. Wells, Differential Analysis on Complex Manifolds. Prentice Hall, 1973.

[Yau77] S. T. Yau, *Calabi's conjecture and some new results in algebraic geometry*, Proc. Nat. Acad. Sci. USA 74 (1977), 1789–1799.

[Zucker79] S. Zucker, *Hodge theory with degenerating coefficients*, Ann. Math. 109 (1979), 415–476.

[Zuo94] K. Zuo, *Factorizations of nonrigid Zariski dense representations of π_1 of projective algebraic manifolds*, Inv. Math. 118 (1994), 37–46.

INDEX

In the formulas $*$ stands for a variable

Abelian by finite group, 0 3 2 3, variety, 0 1 1
alb$_*$, Alb($*$), 0 1 2
Albanese morphism, variety, 0 1 2
automorphic form, 5 2, 16 1

base locus, 5 19
Bergman kernel, 7 1
big, 0 4 2
birational transform, 10 3
Bs$|*|$, 5 19

canonical class, line bundle, 0 4 4
cocycle condition, 5 2
connected rationally, rationally chain, 4 10
cyclic cover, 9 4

degenerate Kahler form, 11 1
dim$_*$, 6 9 3, 6 10 2
discontinuous action, 5 5
discrepancy, 10 1 3
divisor, $\mathbb{Q}$-, 0 4 3

equivalence linear, numerical 0 4 1
essentially a subgroup, 3 1
etale, 0 5 8
$E_*(*)$, 13 3

factor of automorphy flat, trivial, 5 2
free (group action), 5 6
free (linear system), 0 4 9
fundamental domain, 5 6 2, compact, 5 6 3
fundamental group algebraic, 4 2, large, generically large, 4 6, large algebraic, generically large algebraic, 4 6

general point, subvariety, very general point, very general subvariety, 1 4
generically finite, 0 1 3

$\kappa(*)$, 4 11
K_*, 0 5 4
klt, 10 1 5
Kodaira dimension, 4 11

L^2-cohomology, 6 2 2
L^p-metric, 10 5
L^p-section, 5 12

lc, 10 1 6
linear equivalence, 0 4 1
log canonical, 10 1 6
log resolution, 10 3

map, 0 4 7
morphism, 0 4 7

nef, 0 5 2
Neumann, von Neumann dimension, 6 6–10
normal cycle, 2 1
  family of, 2 2
  locally topologically trivial family of, 2 3
  weakly complete locally topologically trivial family of, 2 7
numerical equivalence, 0 4 1

operator positive, Hilbert–Schmidt, 6 7 1

$\pi_1(*)$, $\hat{\pi}_1(*)$, 0 4 6
pluricanonical, divisor, map, 0 4 5
plurigenus, 0 4 5
Poincare series, 5 2, 5 4, 5 10
$P_*(*)$, 0 5 5

$\mathbb{Q}$-Cartier divisor, 0 4 3
$\mathbb{Q}$-divisor, 0 4 3

Res(X), 14 4 1
residually finite, 4 2

sh_*, $Sh(*)$, 1 3
$\underline{\mathrm{sh}}_*$, $\underline{\mathrm{Sh}}(*)$, sh^*_*, $\mathrm{Sh}^*(*)$, 3 5
$\widehat{\mathrm{sh}}_*$, $\widehat{\mathrm{Sh}}(*)$, 4 3
Shafarevich map 3 5, algebraic map, 4 3, morphism 1 3, 3 5, variety, 1 3, 3 5
singular metric, 10 5, 11 1 2
singularity rational, 12 2, Du Bois, 12 7

uniruled, 4 10

$VG(*)$, 2 4

$\| * \|$, 5 12
$\ulcorner * \urcorner$, $\llcorner * \lrcorner$, $\{*\}$, 9 7
$*[\sqrt{*}]$, 9 4
$\sim$, 0 5 1
$\equiv$, 0 5 1

Milton Keynes UK
Ingram Content Group UK Ltd.
UKHW021413010924
447708UK00016B/146